PREPARATIVE TECHNIQUES

METHODOLOGICAL DEVELOPMENTS IN BIOCHEMISTRY

edited by Eric Reid

Wolfson Bioanalytical Centre *University of Surrey*

Other titles in the Series

SEPARATIONS WITH ZONAL ROTORS *(now designated Volume 1)*

published in 1971 by the Wolfson Bioanalytical Centre of the University of Surrey at Guildford

distributed by the University of Surrey Bookshop, Guildford

METHODOLOGICAL DEVELOPMENTS IN BIOCHEMISTRY, Volume 3: ADVANCES WITH ZONAL ROTORS

published by Longman Group Limited

Methodological Developments in Biochemistry

2. Preparative Techniques

Edited by Eric Reid

Wolfson Bioanalytical Centre
University of Surrey

Longman

LONGMAN GROUP LIMITED
London
Associated companies, branches and representatives
throughout the world

First published 1973

ISBN 0582 46010.7

Printed in England
by Lowe and Brydone (Printers) Limited

Editor's Preface

The aim of the series METHODOLOGICAL DEVELOPMENTS IN BIOCHEMISTRY, of which the present book comprises Volume 2, is not merely to reinforce the collections of reference books in libraries. Rather it is to benefit individual bench workers and advanced-level teachers, at a price which they can afford, with desk information on techniques that are valuable in biochemical and related fields. Whereas Volume 1, *Separations with Zonal Rotors*, started almost from first principles, henceforth most of the articles will assume some knowledge of underlying principles and seek to present approaches which are recent yet not ephemeral. The Editor has had gratifying cooperation from contributors, and also from the British Biophysical Society insofar as the present book has arisen from a Symposium held under their auspices. Proofs were not usually sent to authors, who are therefore absolved if typographical errors have crept through.

Whilst the emphasis in the present Volume is on preparative methods, some attention is paid to analytical methods, particularly in Section 21. The development and application of equipment and methods is in fact a major interest of the Wolfson Bioanalytical Centre, coupled with the running of Workshop Courses and Symposia where demand exists; the Centre was launched with a pump-priming grant from the Wolfson Foundation, and lives largely by commissioned work. The present publication venture has hinged on the enthusiastic cooperation of colleagues in the Centre, including Diane Mohr who skilfully typed camera copy. Also vital has been the pleasant collaboration with Longman Group staff.

Permission to adapt published material has kindly been granted by Elsevier-North Holland (Table 1 in Article 11, from *FEBS Letters*, and Figs. 1 & 2 in Article 16, from *Biochim. Biophys. Acta*), by Springer-Verlag (Figs. 3-5 in Article 16, from *Eur. J. Biochem.*), and by the *Biochemical Journal* (Table 1 in Article 13).

Wolfson Bioanalytical Centre
University of Surrey

Eric Reid

10 November 1972

CONTENTS

* *Section 21 includes comments relating particularly to these articles*

Chromatographic and affinity methods, *continued* Page

AUTHOR LIST
for main contributions

For full addresses, see headings to the cited Articles — first look them up in Contents (pp. 6-7), which also lists supplementary contributions whose authors are not listed below.

A. Brownstone - Art. 2
Nat. Inst. for Medical Res., London, U.K.

G.B. Cline - Art. 18 (also 21)
Univ. of Alabama, U.S.A.

O. Cromwell - Art. 13
Univ. of Surrey, U.K.

A.M. Denman - Art. 20
Clinical Res. Centre, Harrow, U.K.

G.N. Dessev - Arts. 5 & 6
Acad. of Sciences, Bulgaria

K.A. Dicke - Art. 19 (also 21)
Radiobiological Inst., Rijswijk, The Netherlands

J.S. Fawcett - Art. 7
Queen Mary Coll., Univ. of London, U.K.

G.P. Georgiev - Art. 15
Acad. of Sciences, U.S.S.R.

K. Grancharov - Art. 5
Acad. of Sciences, Bulgaria

M.E. Haines - Art. 11
Univ. Coll. London, U.K.

A.D.R. Harrison - Art. 3
Univ. of Surrey, U.K.

R.H. Hinton - Art. 6
Univ. of Surrey, U.K.

S. Hjerten - Art. 4 (also 21)
Univ. of Uppsala, Sweden

A.M. Holmes - Art. 11
Univ. Coll. London, U.K.

K.G. Huggins - Art. 12
Miles-Seravac, U.K.

G. Johansson - Art. 17
Univ. of Umeå, Sweden

I.R. Johnston - Art. 11
Univ. Coll. London, U.K.

V.L. Mantieva - Art. 15
Acad. of Sciences, U.S.S.R.

B.J. Miles - Art. 10
A.E.R.E., Harwell, U.K.

C.J.O.R. Morris - Art. 1
Queen Mary Coll., Univ. of London, U.K.

B.K. Pelton - Art. 20
Clinical Res. Centre, U.K.

A. Pusztai - Art. 16 (also 21)
Rowett Inst., Aberdeen, U.K.

E. Reid - Art. 3 (also 21)
Univ. of Surrey, U.K.

L.C. Robinson - Art. 9 (also 21)
Lister Inst., Elstree, U.K.

A.P. Ryskov - Art. 15
Acad. of Sciences, U.S.S.R.

O.P. Samarina - Art. 15
Acad. of Sciences, U.S.S.R.

A.R. Thomson - Art. 10
A.E.R.E., Harwell, U.K.

R. Williamson - Art. 14
Beatson Inst. for Cancer Res., Glasgow, U.K.

1 A NEW CLASSIFICATION OF SEPARATION METHODS

with an addendum on Fractionation of macromolecules in gels

C.J.O.R. Morris
Department of Experimental Biochemistry
London Hospital Medical College at Queen Mary College
Mile End Road, London E1 4NS, U.K.

A classification of separation methods serves a useful purpose insofar as it stimulates the application of existing concepts to new problems, and thus leads to the development of new methods.

Several such systems of classification have been proposed. The best-known is that into zonal, frontal and displacement methods suggested by Tiselius (1) in 1945, and based on the proportion of the separation channel occupied by the initial zone. Other classification systems are based on the nature of the molecular mechanism involved in the separation, and are perhaps less useful in suggesting new methods.

I would like to suggest here an alternative classification based on the fundamental physico-chemical nature of the process which brings about the separation, i.e. whether it is a kinetic or an equilibrium process. Of course all separation methods involve equilibria to some extent, for example between phases in chromatography, and between ionic and non-ionic forms in electrophoresis. The proposed distinction is rather between those methods in which separation occurs as a result of continuing migration, and those in which the separated solutes occupy static positions at equilibrium. The distinction corresponds to a basic division in physical chemistry.

Until quite recently the majority of separation methods were kinetic in character, and depended on differences in the relative rates of migration of the components under specified conditions. These include virtually all chromatographic methods and the majority of electrophoretic and sedimentation separations. Two or more components are impelled along the separation channel, and as a result of their different rates of migration the distance between the zone centres increases both with time and with the distance traversed and is in most cases directly proportional to these variables. The zones also spread out laterally due to various diffusion mechanisms, and this zone dispersion opposes the resolution of adjacent zones. Fortunately this effect is proportional to the square root of the time or distance traversed, so that it increases less rapidly than the separation of the zone centres by differential migration. All kinetic separations are possible only because of this basic relation. Although it would appear that the resolution of difficultly separable components might be improved by prolonged migration, this is only partially true, since zone dispersion may then be so great as to bring the solute concentrations in a zone below practical limits of detection or collection. The important point to emphasize is that all kinetic separations are intrinsically time-dependent so that the success or failure of a particular separation depends critically on the relative rate constants of migration and diffusion.

Entirely different mechanisms are operative in separations based on equilibria. The solute mixture can initially occupy the whole of the separation channel, and during the course of the fractionation the solute molecules will migrate in both directions towards their final equilibrium position. This process continues until an equilibrium is reached between solute concentration by converging migration and solute dispersal by diffusion. This is reflected in the zone profile function -

$$\phi_1 = \exp\left(- \frac{H \cdot dy/dx}{D}\right) \qquad (1)$$

where H is the field strength, D is the diffusion coefficient and dy/dx is the gradient of the separation variable (e.g. pH or density) y with distance x along the channel. It is evident that large negative values of the exponential giving a sharp zone are obtained with large values of the field H and of the gradient dy/dx, and small values of D, and also that equation (1) contains no time-dependent variable. In contrast the zone profile function for a kinetic separation is given by -

$$\phi_2 = \exp\left(- \frac{H \cdot u}{D}\right) \qquad (2)$$

where u is the zone mobility in the system, and is obviously a time-dependent variable.

Time independence confers several important advantages on equilibrium separations. In the first place the experiment can be continued until all the solutes are concentrated in their equilibrium zones, so that high solute concentrations can be obtained within the zone, a particularly important factor for preparative separations. In practice the mass of solute within a zone is limited only by its density stability and by solubility considerations. The solutes can also be introduced as a dilute solution in a large volume, and will be concentrated up to a hundred-fold during the fractionation, in contrast to kinetic separations where the solutes are always recovered in solutions more dilute than the initial solution. A third advantage is that the zone centre corresponds to a specific value of a physical property such as density or isoelectric point, which can be used to characterize the solute.

In order to provide adequate spatial separation between the equilibrium zones it is necessary to provide a gradient dy/dx of the property on which the separation is based, e.g. pH in the case of isoelectric focusing.

The gradient can either be pre-formed or be set up during the course of the fractionation, as in isopycnic banding by the Messelson method or in isoelectric focusing. Both methods give very similar results. It is possible that the gradient could be replaced for certain preparative applications by a stepwise system, with a large increase in zone loading, but the possibility does not appear to have been explored.

Up to the present time equilibrium separations have been confined to isopycnic and isoelectric fractionations in density or pH gradients. The only comparable application in a chromatographic system is the zone precipitation method of Porath (2) in which the solutes move through a preformed gradient of increasing salt concentration until they reach a point at which they are just precipitated. The gradient itself moves through the column so that the precipitated zones can be collected in the effluent. The method retains the important zone-sharpening effect of equilibrium. The principle would appear to be applicable to other solute properties in a chromatographic system, such as isoelectric equilibrium on an ion-exchange column, but these possibilities do not appear to have been tried.

Equilibrium separation methods can be readily adapted for continuous operation in crossed-field systems. Continuous zone removal has been achieved in zonal centrifugation, and a continuous flow system can be used for isolectric focusing (3). A striking advantage of this system compared with the corresponding kinetic electrophoretic system is that the exit position is independent both of the applied potential and of the flow rate, and is therefore inherently stable during long periods. It should be possible to build in comparable stability into any continuous flow equilibrium method.

In conclusion I hope that this classification with its emphasis on the important features of equilibrium methods will stimulate others to apply these principles in new fields.

References

1. Tiselius, A. *The Svedberg Memorial Volume*, Almqvist and Wiksell, Uppsala (1945), p.370.

2. Porath, J. *Nature (Lond) 196* (1962) 47.

3. Fawcett, J.S., *this volume*, p. 61.

Addendum: Fractionation of macromolecules in gels

Macromolecules can be fractionated on molecular sieving media by hydrodynamic (chromatographic) or electrical transport methods. In the former case the separation mechanism is wholly diffusional, while in the latter case both diffusional and electrophoretic mechanisms are operative. The diffusion within the restricted channels of the gel medium is clearly similar in the two cases, and a unified theory has been proposed for these separations (1). This is embodied in the basic equation:

$$\mu = aK_{av} + b$$

where μ is the reduced electrophoretic mobility of the solute in the gel medium, and K_{av} is its partition coefficient in molecular sieve chromaography in the same medium. a is a slope constant approximately equal to one, and b is a small intercept constant. The validity of this relation is completely supported by a considerable amount of experimental evidence from polyacrylamide gels of differing composition, and it appears to be justifiable to extend it to other gel media. The two methods are thus clearly complementary both for preparative applications and for the estimation of molecular dimensions.

Reference

1. Morris, C.J.O.R. and Morris, P., *Biochem. J. 124* (1971) 517.

2 PREPARATIVE ACRYLAMIDE GEL ELECTROPHORESIS

A.D. Brownstone
National Institute for Medical Research
Mill Hill
London NW7 1AA

This article represents an account of the 'state-of-the-art', from a practical viewpoint.

The use of polyacrylamide gel for molecular sieve electrophoresis on an analytical scale affords a high resolving power under a wide variety of conditions. The development of the method for use on a preparative scale poses a number of problems. The purpose of this article is to discuss these problems from a practical point of view.

I. TYPES OF APPARATUS

The design of apparatus for preparative polyacrylamide gel electrophoresis follows broadly that used for analytical work; the gel is cast either in the form of a flat relatively thin slab, or as a cylinder. As the size of the apparatus increases from that suitable for separating up to about 10 mg of protein material to that used for one or two grams, difficulties arise such as those associated with casting the gel, heat dissipation, and collecting the separated zones. Experience has shown that the set-up should be as simple as possible in order to avoid frustration arising from everyday handling and assembly. Some designs more closely approach this ideal than others.

(a) Horizontal slab

The horizontal slab poses difficulties associated with loading the material; this is done by casting a slot in the gel into which the solution is placed. Unless the material is loaded in either dilute polyacrylamide or agarose gel, decantation of the material at the vertical entry face of the running gel will take place, giving rise to band distortion and streaming. Band distortion is also liable to occur due to uneven voltage gradients round the slot, particularly at the corners. The design by Hodson and Latner (1) entails a discontinuous collection of the zones after end elution: this helps to overcome decantation, which also occurs at the vertical membrane used to prevent migration of the material into the electrode chamber. The main advantage of the horizontal gel slab is that by having a collection system at both the anodic and cathodic ends, a mixture containing both positively and negatively charged material may be resolved simultaneously. It is, of course, essential that both upper and lower surfaces of the slab be maintained at the same temperature to avoid serious band distortion (2).

(b) Vertical slab

As with the horizontal slab, this type of gel is easily cooled, and provided that the thickness of the slab does not exceed about 2 cm, very little band distortion due to heating should occur. However, practical difficulties often occur in constructing the apparatus if it is desired to take advantage of the better heat conductivity of glass compared with that of plastic material. Unlike the horizontal slab the loaded volume is not limited.

(c) Cylindrical gels

The majority of designs for preparative scale apparatus favour casting the gel in the form of a cylinder with or without a central cooling finger (3-13). This is partially due to the relative ease of construction from readily available tubing of different sizes. Moreover, gels having a large surface area may be contained in a compact apparatus.

II. CONSIDERATIONS OF DESIGN AND USE

(a) Composition and casting of the gel

It is recommended that the materials used for making the gel be of high quality. Great care must be taken in handling acrylamide and N,N-methylene bisacrylamide (BIS) owing to their toxic nature (14). The conductivity of the gel monomer solutions before addition of buffer should not differ from that of the deionized water in which they are dissolved. If the solution is found to have a higher conductivity it should be treated with a mixed cation/anion exchange resin or the solid monomers should be recrystallized (acrylamide *ex* chloroform, BIS *ex* acetone (15)). Recrystallized materials should be stored at 2°C. If these precautions are not taken it will quite often be found that the gel shrinks considerably when subjected to electrophoresis even in the absence of protein. The gel should be prepared from the solid monomers rather than stock solutions; the latter tend to deteriorate on storage, giving rise to an unduly high A_{280} background of the eluting buffer (2). High background readings will also be obtained if the gel is not left a sufficiently long time for polymerisation to be completed before it is used. At least 4 hours at room temperature or overnight at 5°C should be allowed. However carefully the gel is prepared it is always found that the A_{280} of the collecting buffer is increased after its passage across the gel.* When the gel is cast in the apparatus in which it is to be used, the temperature of the monomer solution should be the same as that at which it is intended to run the electrophoresis.

It is well known that acrylamide-BIS solution will not polymerize

* *See Article 3 in this book, by A.D.R. Harrison & E. Reid, which concerns UV-absorbing impurities in gel eluates. - Ed.*

in contact with air, and in order to protect the exposed surface of the monomer solution it is usual to overlay it with water. As the surface area of the gel is increased it becomes more difficult to ensure a perfectly flat gel surface normal to the direction of migration when over-layering with water is performed.

(b) Mould for casting the gel

The use of a separate mould in which to cast gels having surface area greater than 30 cm^2 is strongly recommended (7). The only limitation to the use of a mould is where the gel must adhere to the walls of the running apparatus to prevent it falling into the collecting chamber. In large preparative apparatus with fixed volume elution chambers, it is essential to support the gel; this is usually accomplished by arranging for the gel to rest on a pad of porous material such as sintered glass or porous plastic (4,6,13). Whatever material is chosen, attention must be given to the possible electrodosmotic flow of buffer through the pad which occurs if the pores are too fine.

The design of a mould for cylindrical gel should incorporate the following features: (1) It should be constructed of Perspex or other similar transparent plastic to which the gel does not adhere; (2) the end plates should be flat and easily removable from the cylindrical section; (3) The seal between the end plates and the cylindrical section must be leak-proof; (4) the mould is filled through a hole in the cylindrical wall, a small reservoir being fitted on the outside at this point to exclude air; (5) if provision is made for a central cooling finger, the hole for this is cast by a suitably sized rod positioned concentrically in the mould and fixed to one of the end plates; (6) the dimensions of the mould should be as follows: a) length as desired; b) overall diameter 5-7% larger than that of the running apparatus; c) diameter of the hole for central cooling finger 5-7% less than that of the finger itself. By casting the gel slightly larger than the running apparatus, leakage between the gel and the walls is completely eliminated unless gross shrinkage takes place during the electrophoresis.

No difficulty should be found in loading cast gels of from 3.5% to 7% total monomer content. More dilute gels may require to be stiffened slightly with agarose (see below); more concentrated gels (up to about 20%) may require the use of a simple bevelled guide.

(c) Polymerization catalysts

Large gels for preparative work cannot be satisfactorily polymerized by the use of light-activated riboflavin; the most usual and satisfactory procedure is the use of ammonium or potassium persulphate in conjunction with N,N,N',N'-tetramethyltheylene diamine (TEMED). The amounts of the catalyst and base to be used will depend on a number of factors such as a) temperature of monomer solution, b) pH of monomer solution, c) whether the monomer solution has been degassed or not, d) concentration of monomer solution.

At room temperature and above pH 5 in non-degassed solutions a gel will form in 6-10 min if 0.1% w/v ammonium persulphate and 0.1% v/v TEMED. are used. If the solutions are degassed the amounts of persulphate and TEMED may be reduced to 0.05% w/v and 0.05% w/v respectively for a gelation time of 6 min. Degassing is *not* recommended if the gel is to be cast at pH above 5 in a mould. The smaller quanitity of persulphate required causes the monomer solution to be sensitive to the slightest contact with air, so that if one or two quite small air bubbles are trapped in the mould a relatively large volume of solution surrounding them will not polymerize and the resulting block of gel may be rendered useless. For gels polymerized at low temperature (2-10%), up to 0.3% persulphate and TEMED are required if a polymerization time of 15 min is required with a non-degassed solution. Special care must be taken if volumes of 30 ml or more of gels containing 20% or more of monomer are to be used, as the amount of heat evolved may well heat the gel to over 100° and since maximum heat evolution seems to take place just after the gel has set: the gel will break up and in extreme cases it may explode. For acid gels, i.e. below pH 4, the alternative catalyst system of Jordan and Raymond (16) may be used. This replaces persulphate and TEMED. with 0.1% ascorbic acid, 0.0025% ferrous sulphate and 0.003% H_2O_2.

(d) Use of agarose

Very dilute polyacrylamide gels, i.e. below 3% w/v total monomer, may be mechanically stiffened by the use of agarose (17). The amount to be used should not exceed 0.3% w/v if the gel is to be cast in a mould, otherwise it will be too brittle to handle. When the gel is cast in the apparatus the quantity of agarose is not critical and may be up to 0.7% w/v. The agarose is dissolved in about two-thirds of the final gel volume of water or buffer solution by heating the suspension on a boiling water bath until it has totally dissolved and no discrete particles are visible. The solution is then cooled to 40° and the solid acrylamide, BIS and persulphate are added and dissolved; water at 40° is added to the required final volume. Finally the TEMED is added with gentle stirring and the solution is poured into the apparatus or mould where it is left for some hours for setting and polymerization to take place. The amounts of persulphate and TEMED should not exceed 0.08% w/v and 0.8% v/v respectively to avoid premature gelling of the warm acrylamide solution.

(e) Proportion of BIS to acrylamide

The variation in the properties of the gel caused by altering the proportion of BIS has been studied by a number of workers (18 and, particularly pertinent for electrophoretic work, 19 & 20). The composition of the gel, i.e. total monomer concentration, should be determined by prior analytical-scale experiments. However, it should be remembered that with the end elution type of apparatus all bands must travel the total length of the gel. Thus if it is desired to collect the slow-moving bands in a particular mixture it may be possible to use a slightly lower monomer content to obtain a good resolution in a reasonable time.

III. BUFFER SOLUTIONS

The buffer solution to be used will depend on the material to be separated, and preliminary analytical experiments should be made to determine the optimum conditions for the preparative experiment. Details of suitable buffer solutions for electrophoresis have been published (e.g. 2, 21-23). Some points to be taken into consideration are now given.

(a) Conductivity of the buffer solution

This should be as low as possible in order to achieve a high potential gradient across the gel without the production of excessive heat.

(b) Choice of buffer

In order to keep the volume of buffer solution required within reasonable limits, most designs of the larger types of preparative apparatus employ a recirculating buffer system. This ensures that conditions of pH and concentration are held constant in both anode and cathode chambers. Buffers used in this type of system must therefore be stable, as any decomposition products formed at either electrode will also be continuously recirculated and may harm the material being run. For this reason alkaline buffers containing chloride ions should be avoided because of electrolytic formation of hypochlorite. The conductivity of Tris/glycine buffer has been found to increase at a rate dependent on the amount of current flowing, and when used on a large scale the tank buffer must be changed every 6-8 h. (In a specific example, with 8 l of 0.015 M Tris/0.05 M glycine at pH 8.7 the starting current was 300 mA at 180 V; after 8 h the current had increased to 400 mA at 170 V although the pH remained constant.)

(c) Discontinuous buffer systems

To obtain a sharp starting zone, the use of discontinuous buffer systems such as those developed by Ornstein (24) is sometimes recommended. However, except in special circumstances (see end of article, *Recycling*) it is usually preferable to use a concentration gradient if it is desired to form a sharp starting zone on a preparative scale (25).

(d) Care in loading protein in buffer having high conductivity in dilute solution

When using such buffer systems, the high conductivity necessitates use of a low voltage (25-40 V) between the electrodes to avoid excessive heating. This in turn means that the material may take one hour or more to move into the gel, by which time the pH in the cathode compartment of an apparatus designed for buffer recirculation may well have changed considerably owing to depletion of ions. This may be avoided if a baffle is inserted between the electrode and gel so that recirculation can be maintained round the electrodes without disturbing the layer of protein solution. The

The depletion rate of ions from the chamber below the gel should be calculated by Faraday's Law from the current and the volume and concentration of the buffer.

(e) Use of sodium dodecyl sulphate (SDS) and/or urea

The use of SDS presents no difficulties when used in an apparatus designed for buffer recirculation. However, to prevent its crystallisation the tank temperature must not be allowed to drop below about 12-15° for solutions containing up to 0.1% w/v. The concentration of SDS in the tank and collecting buffer must be the same as that in the gel. In recirculating systems urea must not be allowed to come into contact with the electrodes (4,7); if this happens it will be found that the A_{280} of the buffer throughout the system will increase rapidly, rendering protein detection by this method impossible. This problem may be overcome by use of a suitable baffle system (7) which has enabled runs of 24 h to be completed without trouble. The tank temperature must also be adjusted to avoid crystallisation of the urea. In order to prevent the gel swelling or shrinking due to osmosis, the urea concentration of the solution in immediate contact with the gel must be kept very close to the original concentration in the gel.

IV. LOADING THE MIXTURE TO BE RUN

The following section applies particularly to apparatus in which the material is loaded onto the upper horizontal surface of the gel block. In the most usual procedure the mixture of substances to be run is dissolved in a suitable buffer solution to which has been added up to 5% w/v sucrose or 2% glycerol. The resulting solution is then loaded onto the gel surface where it forms a sharp layer under the tank buffer.

(a) Quantity of material to be loaded

The approximate load of any particular mixture per unit area of gel surface should be decided by conducting preliminary runs on analytical gels (7). Since the width of the bands depends on the amount of substance they contain, the quantity of a mixture which may be loaded will be limited by the proximity of adjacent zones as they emerge from the gel. It is therefore desirable that as much preliminary purification as possible be carried out before gel electrophoresis. The presence in the mixture of relatively small amounts of slow-moving substances is also a limiting factor to the load which may be applied. The slow-moving material tends to block the surface of the gel, thus preventing the entry of the bulk of the mixture which then forms a highly viscous thin layer on or just under the gel surface. It is sometimes possible to overcome this difficulty by repeatedly reversing the polarity of the electrodes. Thus, the polarity is reversed for ½-1 min causing the blocking substance to lift from the gel surface; on reapplying the normal polarity for 2 or 3 min more of the mixture will enter before the surface becomes blocked again. Finally, after the last polarity reversal the power supply is turned off and the surface of the gel is gently scraped to remove any adhering material. The contaminated buffer solution above the gel should then be removed and replaced before the run is continued. It is good practice to clean the surface of the gel, as in the final stage above, in any run. In this way excessive and premature shrinkage of the upper part of the gel may be avoided.

In order to prevent blockage of the gel surface by large components in a mixture, these may sometimes be 'trapped' by loading the mixture in an agarose gel. This is poured over the surface of the running gel, allowed to set and then the tank buffer carefully added. When all the migrating substances have been run out of the agarose into the acrylamide gel it is essential to remove the agarose layer completely. Failure to do this may lead to shrinkage of the gel and distortion of the separating bands.

The gel may also become blocked by overloading at the surface due to use of zone-sharpening techniques. This is particularly liable to happen with very large molecules, e.g. thyroglobulin, or with non-globular molecules. By loading such molecules in more dilute solution or more usually in more concentrated buffer solutions a broader starting zone is obtained (7). Accidental zone sharpening can occur if the mixture is dissolved in a buffer solution capable of producing a 'front' between fast and slow moving ions such as occurs in the 'disc' analytical procedure. This may happen due to incomplete removal of such ions, e.g. phosphate or chloride, when pre-dialysing into glycine or borate running buffers.

V. BAND DISTORTION

Distortion of the separating bands is a major factor affecting the resolution of mixtures during preparative gel electrophoresis. It may be caused by a number of factors, the most common of which are discussed below.

(a) Heating effects

The mobility of a charged ion moving in an electric field is inversely proportional to the viscosity of the medium in which it moves. Since viscosity is temperature dependent, it follows that mobility is also affected. Morris and Morris (26) give an approximate figure of 2.5% increase in mobility per degree rise in temperature. Ohmic heating of the gel block during electrophoresis is unavoidable, and therefore good design of the cooling system is crucial if distortion due to temperature gradients is to be minimised. The usual method of cooling the gel from the longitudinal sides, either of a slab or cylinder, is bound to give rise to a temperature gradient across the path of ghe moving bands. If the thickness of the gel between the cooling surfaces does not exceed 2 cm the temperature gradient may be held to 5° and distortion is not excessive. It follows that when a cooling finger is used in a cylindrical gel, the greatest care must be taken to see that it is mounted centrally, otherwise the bands will run faster where the distance between the cooling surfaces is greatest.

For large blocks of gel (capable of resolving 500 mg and over), where the surface area on which the mixture is loaded exceeds 60 cm^2, it is better to cast the gel as a cylindrical block, with no central cooling finger, and to arrange for the block to be cooled only from the ends. The wall of the gel holding tube is insulated, minimising the radial temperature gradient and producing a longitudinal temperature gradient in the gel. This means that minimal longitudinal distortion of the band takes place as it moves through the gel. This system involves a higher temperature halfway down the gel than would be the case if it were possible to use efficient longitudinal cooling. For example, the temperature 2.5 cm from the upper surface of a cylindrical gel block 9 cm diameter x 5 cm long with recirculated tank buffer at 2° was 14° with 40 watts being dissipated between the electrodes (200 mA, 200 V). The radial gradient at this depth was 4°. If the length of the gel exceeds 6 cm then the current must be limited if the temperature at the centre of the gel is not to exceed 20°.

(b) Electrodes

The electrodes must be shaped to provide an even voltage gradient over the whole area of the gel. They are usually made from platinum wire although ideally platinum gauze would be better. For cylindrical gels the wire should be mounted on a plastic ring, the diameter of which is not less than 90% of the diameter of the gel. The wire should be fixed round the circumference and also across the ring to form an open mesh. The two electrodes must be supported parallel to each other and to the ends of the gel. They should not be mounted closer than 3 cm to the ends of the gel to avoid band distortion. When using a buffer with high conductivity, which therefore passes a high current, there will be a considerable voltage drop between the electrode and buffer solution due to the insulating effect of the gas being produced. This voltage drop can be minimized by using more wire to increase the surface area of the electrode.

(c) Uneven loading

Serious band distortion and trailing will result if the gel is overloaded with respect to any one component in the mixture. Even if the gel surface is not blocked completely it may be found that a highly concentrated zone is formed just under the surface from which the material streams. This may also happen if the running buffer concentration is too low.

Local overloading of the surface may occur if it is not horizontal during loading. Also, if the conductivity of the buffer in which the mixture is loaded differs from that of the tank and gel buffer, an uneven voltage gradient across the layer will cause band distortion if the layer is not of even depth over the gel.

VI. RECOVERY OF THE SEPARATED ZONES

This is achieved by one of two general methods.

(a) Segmentation of the gel

This method is limited in practice to gels in the form of a slab. Electrophoresis is continued only until the fastest moving required band has migrated almost to the end of the slab. The bands are located by staining one or more small sections of the gel, which is then cut into segments containing the required zone. The material is recovered from the segments by a further stage of electrophoresis into a relatively small volume of buffer (27,28) or into a sucrose/salt gradient in which it forms a concentratrated zone (29). In the latter case the gradient has to be fractionated and the sucrose and salt removed from the required material by a suitable method.

(b) End elution during electrophoresis

This method may be used either with slab (1,30) or cylindrical gels. The electrophoresis is continued until the slowest required component has migrated through the whole of the gel. On emerging the material is collected in either a continuous (3-6) or an intermittent (1,7-9) flow of buffer. The design may incorporate a membrane to prevent the material escaping from the collecting chamber, or the use of a membrane may be avoided (6,29). When using a membrane, consideration must be given to pH changes adjacent to the membrane caused by the Bethe-Toropoff effect (33). Hjertén *et al.* report that coloured proteins lose or change their colours on coming into contact with the membrane and may be denatured (30). Many other authors have *not* reported trouble from this cause. Continuous buffer flow collecting systems employ a fixed volume chamber, under the gel, through which the collecting buffer flows. In such designs some means of preventing the gel from falling or swelling into the chamber must be provided. In order to obtain efficient removal of the emerging bands, the volume of the chamber must be as low as possible and the material must be removed from the whole area of the gel with equal efficiency and in the minimum volume of buffer. This becomes progressively more difficult to accomplish as the size of the apparatus increases. Some designs arrange for the buffer to flow from the circumference of the chamber radially to the centre where it is removed. Hjertén *et al.* (6,30) avoid the use of a dialysis membrane and report increased efficiency by filling the chamber with Sepharose 4B. They have also published a design for a large slab gel apparatus using this system (30). In all these designs the flow rate of the buffer stream must be fast enough to overcome the electrophoretic migration of the required molecules. If the flow is too slow there will be a low recovery of material, which will be held onto the membrane by the electric field, or, in the case of membraneless designs, will migrate into the main buffer tank. The relatively high flow rate required leads to considerable dilution of the collected material, and minor bands may be very difficult to detect by the usual UV monitoring apparatus.

Intermittent collecting systems

In the simplest of these (8) the emerging material is trapped in a dialysis sac which is fixed to the lower end of the gel tube. The sac has to be changed in order to prevent re-mixing of the bands. A carefully timed study of the pattern obtained on an analytical gel run under identical conditions is claimed to give the required interval between sac changes. By using a simple timing mechanism to remove and replace the collecting buffer at pre-set intervals the process is made more automatic (9).

All the above designs are fixed-volume collection chambers with attendant problems of gel support. An apparatus designed by the author (7) uses a variable volume collection chamber with intermittent collection. This enables an extremely simple and trouble-free design of gel holder to be used. The pre-cast cylindrical gel is located in the lower end of a tube, the wall of which is insulated. The gel is supported by a dialysis membrane stretched across the lower end of the tube and held in place by a tight-fitting ring. The collecting buffer is pumped between the gel and the membrane where it spreads out to form a thin layer. After a pre-set time interval it is all pumped out until the gel and membrane are in close contact and a fresh fraction of buffer is pumped out. The associated control system enables the interval between removing and replenishing the collecting buffer to be varied automatically from 7 to 45 min as the run proceeds. Each time the collecting buffer is removed the electrophoresis power supply is first reversed for about 15 sec and then turned off until the fresh buffer is pumped in. This prevents the material being held onto the membrane by the electric field. Gels of 6, 9 and 15 cm diameter have been used. By using intermittent collection the overall flow rate is considerably reduced compared with the continuous flow procedure. For example, each fraction volume from a 9 cm diameter gel is 2.5 ml, and at the most frequent collecting rate this represents only 17-18 ml per h. This means that in some circumstances a band may be collected in a smaller volume of buffer than the loading volume. No deformation of the lower surface of the gel has been found in this apparatus.

Membranes

Cellophane dialysis membrane is used in many designs to prevent the separated material escaping from the collecting chamber. The permeability of this material to large molecules varies according to the type used (31), and loss of molecules of molecular weight below 20,000 can occur if no precautions are taken, particularly in intermittent collection systems: e.g. cytochrome *c*, with a molecular weight of 13,000, will penetrate dialysis membranes under electrophoresis in a matter of 10-15 min in a field of 10 V/cm.

The membranes may be made relatively impermeable by acetylation (31-32). When treated in a mixture of 50 parts acetic anhydride and 50 parts pyridine at 60° for 12 h (32) they become almost impermeable to bromophenol blue. After soaking in this mixture the membrane becomes very soft and flexible. Immediately before attaching to the apparatus, a membrane treated in this way should be given a quick rinse in water to remove the reagents on the surface. When finally in position it should then be soaked in dilute (about 0.1N) acetic acid until there is no smell of pyridine. By this time the membrane will be hard and tough. Treatment with a mixture of 40 parts acetic anhydride and 60 parts pyridine overnight at room temperature produces a slightly less 'tight' membrane which remains reasonably flexible after removal of the acetylating reagents.

The negative charge on cellulose type membranes means that under alkaline conditions there will be an electroendosmotic flow of fluid through the membrane from anode to cathode. Also the conductivity of the collecting buffer emerging from the apparatus will be higher than its initial value. In runs lasting more than about 12 h it may be found that a zone of increasing conductivity spreads up the gel from the end nearest the membrane. This causes the bands to slow up as they reach this point of the gel, and they may take longer to emerge from the gel than would be expected from studies on analytical gels.

Under acid conditions when running protein from anode to cathode the reverse of the above effects will take place, and in an intermittent collection apparatus having small collecting chamber volume the buffer ions may become exhausted, leading to a dramatic decrease in electrophoresis current; loss of fluid may distort the gel and cause it to stick to the membrane. This effect may be overcome by using a larger volume of more concentrated buffer.

Neutralisation of the membrane charge

Some preliminary work has indicated that the charge on cellophane membranes may be 'neutralized' by treatment with diethylaminoethyl chloride.HCl, thus avoiding the effects described above. It is hoped, in the near future, to publish full details of this work and a recycling process, together with some modifications to the preparative apparatus described in ref. (7).

Recycling

As mentioned above, it is inadvisable to use gels more than about 6 cm long when the gel is cooled from the ends only. When it is desired to separate two adjacent bands which are still not properly resolved after passing through this length of gel, it is possible to re-run the partially resolved bands. An entirely automatic re-loading system has been developed with which, by using a collecting buffer having an anion which will give a zone-sharpening front with the gel and tank buffer, the bands are re-concentrated each time they are re-cycled. Thus the diffusion of the band after three or more passes need be no greater than after one pass. In order to maintain constant conductivity throughout the gel a neutralized membrane (see above) is used.

References

1. Hodson, A.W., and Latner, A.L., *Anal. Biochem. 41* (1971) 522.
2. Gordon, A.H., in *Laboratory Techniques in Biochemistry and Molecular Biology* (T.S. & E. Work, eds.), North Holland, Amsterdam (1969) p.1.

3. Jovin, T., Chrambach, A. and Naughton, M.A., *Anal. Biochem. 9* (1964) 351.

4. Duesberg, P.H., and Ruekert, R.R., *Anal. Biochem. 11* (1965) 342.

5. Gordon, A.H., and Louis, L.N., *Anal. Biochem. 21* (1967) 190.

6. Hjertén, S., Jerstedt, S. and Tiselius, A., *Anal. Biochem. 11* (1965) 211.

7. Brownstone, A.D., *Anal. Biochem. 27* (1969) 25.

8. Kawata, H., Chase, M.W., Elyjiw, R. and Machek, E., *Anal. Biochem. 39* (1971) 93.

9. Schenkein, I., Levy, M. and Weis, P., *Anal. Biochem. 25* (1968) 387.

10. Popescu, M., Lazarus, L.H. and Goldblum, N., *Anal. Biochem. 40* (1971) 247.

11. Coy, P.H. and Wuu, T., *Anal. Biochem. 44* (1971) 174.

12. Altschul, A.M., Evans, W.J., Carney, W.B., McCourtney, E.J., and Brown, H.P., *Life Sci. 3* (1964) 611.

13. Radhakrishnamurphy, B., Dalferes, E.R., and Berenson, G.S., *Biochim. Biophys. Acta 107* (1965) 380.

16. Jordan, E.M. and Raymond, S., *Anal. Biochem. 27* (1969) 205.

17. Dahlberg, A.E., Dingman, C.W., and Peacock, A.C., *J. Mol. Biol. 41* (1969) 139.

18. Fawcett, J.S. and Morris, C.J.O.R., *Separation Sci. 1* (1966) 9.

19. Richards, E.G. and Gratzer, W.B., *Anal. Biochem. 12* (1965) 452.

20. Blattler, D.P., Garner, F., Van Slyke, K. and Bradley, A., *J. Chromatog. 64* (1972) 147.

21. Chrambach, A., and Rodbard, D., *Science (Wash.) 172* (1971) 440.

22. Orr, M.D., Blakeley, R.L. and Panågou, D., *Anal. Biochem. 45* (1971)68

23. Dawson, R.M.C., Elliott, D.C., Elliott, W.H., & Jones, K.M. (eds.) *Data for Biochemical Research, 2nd edition,* Oxford University Press, Oxford (1969), p.504.

24. Ornstein, L., *Ann. N.Y. Acad. Sci. 121* (1964) 321.

25. Hjertén, S., Jerstedt, S., and Tiselius, A., *Anal. Biochem. 11* (1965) 219.

26. Morris, C.J.O.R. & Morris, P., *Separation Methods in Biochemistry,* Pitman, London (1964).

27. Gordon, A.H., *Biochem. J. 82* (1962) 531.

28. Sulizeanu, D. and Goldman, W.F., *Nature (Lond.) 208* (1965) 1120.

29. Hjertén, S., *Biochim. Biophys. Acta 237* (1971) 395.

30. Hjertén, S., Jerstedt, S. and Tiselius, A., *Anal. Biochem. 27* (1969) 108.

31. Craig, L.C., and Konigsberg, W.H., *J. Phys. Chem. 65* (1961) 166.

32. Wink, H., *J. Polymer Sci. 4 [Pt. A-2]* (1966) 830.

33. Svensson, H., *Adv. Prot. Chem. 4* (1948) 251.

3 POLYACRYLAMIDE-LIKE MATERIAL AS A POSSIBLE CONTAMINANT IN PREPARATIVE GEL ELECTROPHORESIS

A.D.R. Harrison and E. Reid
Wolfson Bioanalytical Centre
University of Surrey
Guildford
Surrey, U.K.

We have confirmed and investigated a phenomenon touched on by some authors [e.g. 1-3], viz. the presence of a troublesome contaminant in the effluents from polyacrylamide gel runs. The contaminant seems to be a lyophilic sol, even the sign of the charge being variable. Its ninhydrin colour and its UV absorption spectrum are likewise variable:- there is always some absorption in the 200-210 nm region, whereby the presence of the material can be detected, but in the 260-280 nm region the absorption is often negligible. If, therefore, macromolecules are being separated with monitoring at 260 or 280 nm, a low baseline between the peaks due to the products is no guarantee that the contaminant is absent. No reliable way has been found to obviate the formation and appearance of this contaminant although pre-running for at least 4 h before loading the sample may be of some help. Nor is it easily separable from macromolecular products; its size as judged by ultrafiltration behaviour overlaps with the lower end of the protein range, and when run on Sephadex columns along with protein it tends to accompany the protein. If one cannot find a way to remove the contaminant in a particular macromolecular separation, one must learn to 'live with it'. Analytical evidence suggests that the material may be 'oligo-acrylamide'. Whilst it may vitiate analytical work on separated products, it may be innocuous in test systems as exemplified by alcohol dehydrogenase assay.

In preparative polyacrylamide gel electrophoresis there is an important phenomenon which can be disconcerting to any worker who is unaware of it, viz. the liberation of a material here termed 'X' from the gel itself. Gordon (4) mentions this adventitious material and gives a UV spectrum, together with an A_{280} time course to show its appearance in the electrophoretic effluent (in the context of the use of old solutions for the polymerization).- A peak appeared after 4 h and a second at 15 h: the baseline remained above 0.05. This unwanted material, of which Hjertén (1) was aware, has also been encountered by others (2-8). Thus Winters *et al.* (6) refer to "non-protein materials presumed to be acrylamide polymers of various sizes", and Kaltschmidt and Wittmann (8) used Millipore filters to remove "polyacrylamide particles" from protein products isolated by final batchwise elution of portions of the gel (rather than by continuous elution throughout the run itself. With analytical gels, Loening (9) found that UV-opacity could be obviated by use of recrystallized materials. No author, however, has reported a systematic study of the contamination problem.

PREPARATION AND TREATMENT OF GELS

In most of our own studies of substance 'X', a leaching rather than an electrophoretic procedure has been used for convenience; the material thus obtained seems to be comparable to that found in electrophoretic effluents. The results have been derived mainly from 5% or 10% gels made from 'Cyanogum 41' (American Cyanamid Co., distributed by BDH Chemicals) with 'TEMED' accelerator. As the catalyst we favoured potassium persulphate (10) rather than ammonium persulphate, for the sake of minimizing the risk of appearance of ammonia in the eluate. We have not tried photopolymerization in the presence of riboflavin, since the latter vitiates UV absorption readings such as were made in the following study of 'X', and is not readily washed out. To minimize UV interference by the accelerator (see below), we routinely cut up and washed the gel before preparing test leaches

The fresh gel (usually prepared from freshly dissolved starting materials) was cut up into small pieces, washed and extracted, usually at room temperature, with overnight standing; decantation yielded an optically clear solution. Spectrophotometric readings were made in a SP 1800 instrument (Pye Unicam), routinely including determinations of A_{210} values as a measure of 'X'.

Efforts to minimize the amount of 'X', without any consistent success, have entailed trial of the following variations.-

(a) Different commercial suppliers of the materials used for polymerization have been tried including Serva (Heidelberg; their acrylamide is free of acrylic acid and is kept under nitrogen) and Bio-Rad Laboratories.

(b) The polymerized gel was sometimes treated with methanol, and the 'precipitate' washed with chloroform (which dissolves acrylamide) and with methanol, and taken up in water for repetitions of the treatment; the powder finally obtained was 're-constituted' in water for the electrophoretic run.

(c) The recystallisation procedure of Loening (9) has been tried with the acrylamide and the bisacrylamide.

(d) The proportion of 'BIS' has been varied - this determining the degree of cross-linking (4).

(e) Polymerization temperature was varied between 2° and 75°.

(f) Ageing effects were examined, both with the solutions used for polymerization (no ageing effect was found, contrary to Gordon (4,11)) and with the gel itself. - The leaching procedure gave UV-absorbing material even after 4 weeks, whether or not the gel had already been subjected to repeated leaching.

FEATURES OF 'X' IN RELATION TO DETECTION AND TO EXAMINING PEAKS FROM SEPARATIONS

Spectra

For detecting 'X' we have found A_{210} measurements to be of help. (The term 'A', denoting absorbance, is a valid one, since we have found no evidence of loss of incident light by scattering rather than by absorption.) No reliance can be placed on the A_{280} value as measured by Gordon (4) since this varies from batch to batch, whilst always being comparatively low. Thus in Fig. 1 it will be noted that when two supposedly comparable gels were leached side-by-side, the absorbances of the fresh ('original') leaches in the 260-280 nm region were different relative to those in the 200-210 nm region where the peak was found; the peak position itself was different for the two gels (approx. 200 nm and 209 nm respectively). Fig. 1 further shows that storage of a leach at 7° may diminish the already low A_{280} value, (as likewise encountered with storage at -17°); but this effect of storage is not consistent. There was an apparent rise in a low A_{280} value, with a shift in the absorption peak from 207 nm to near 200 nm, when a leach was dried down and re-constituted; a contributory factor in the latter shift may have been volatilization ('steam distillation') of any residual accelerator during the drying down. Polymerization conditions seem to have no bearing on the spectral variability.

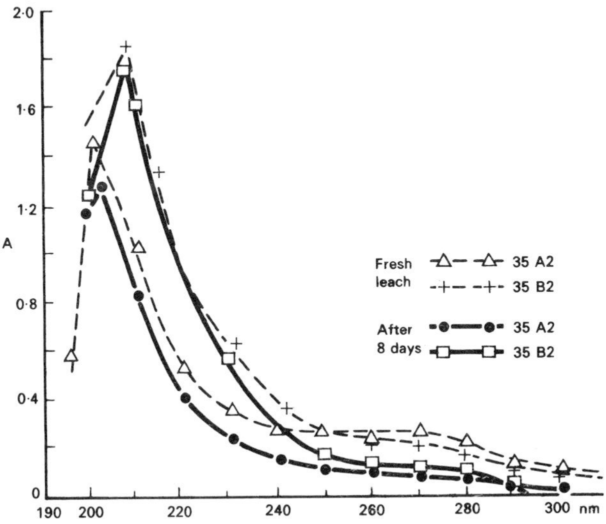

Fig. 1. Absorption spectrum of samples of 'X' derived by leaching two freshly prepared gels (10%), prepared side-by-side, with 12.5 mM borate-HCl buffer pH 8.5. The leaches were re-examined after storage at at 7° for 8 days. *The gels were cut into small cubes (0.5-1 ml) and the first leach (approx. 18 h in the cold, with an occasional stir) was rejected since it would be relatively rich in accelerator and catalyst. The second leach (soak for 18 h) was taken, and filtered before having its spectrum examined. The shapes of the curves should be compared rather than the absolute values for absorbance.*

Evidently the UV absorption behaviour of 'X' is capricious, in accordance with the supposition that 'X' is a lyophilic sol. Yet measurements at 210 nm (205 nm might be preferable) do help in detecting 'X', and up to A_{210} = 1.5 (but not beyond 1.8), Beer's law is obeyed. Detection must not rely on A_{280} readings. The A_{280} value and likewise the A_{260} value, for 'X' in electrophoretic eluates might be deceptively low, as in leaches, and indeed could be swamped by the contribution from any protein (or nucleic acid) that may be present. Insofar as protein also absorbs in the region of 210 nm, as do a variety of molecules, there can be analytical troubles also; indeed there are protein estimation methods which entail UV measurements in this region (12). If A_{210} readings are not made against an appropriate buffer blank, there may be an over-estimate. Examples of A_{210} values for 0.1M buffers are 0.13 for borate (pH 8.8), 1.86 for Tris-HCl (pH 8.8), 0.08 for K_2HPO_4, and 0.01 for KH_2PO_4.

The net A_{210} value due to 'X' in actual electrophoretic runs before loading the sample is typically >1.0 at the outset; even with pre-running overnight it does not fall as sharply as the A_{280} value. Our 'know-your-enemy' findings are now summarized, one problem being 'A_{210} non-additivity'.

Extent of UV and ninhydrin interference by the contaminant

It is important to know the extent to which electrophoretic effluents are contaminated with 'X' on a weight basis, and what interference might be expected if eluted protein is to be analyzed for amino acids. Table 1 gives relevant data, for leaches, and also A_{280}/A_{210} ratios as referred to above. (The tabulated information for substances other than 'X' will be considered later.) It will be noted that the weight of 'X' corresponding to a given A_{210} reading varies 2-fold. Whilst consistent A_{210} readings, conforming to Beer's law, are obtained with a given leach, the application of separation procedures as considered below may show over-recovery of A_{210} units amongst the 'X'-containing fractions. This non-additivity, and the batch-to-batch variability in A_{210} units per unit weight of 'X', suggest that A_{210} varies due to subtle differences in physical state and/or to partial depolymerisation.

It is evident that if in an electrophoretic run the A_{210} baseline (or, less reliably, the A_{280} baseline) due to 'X' were comparable in magnitude to the absorbance due to a protein peak, the weight of 'X' per ml would be similar to that of the protein.

There is an apparent 'crumb of comfort' in that the contaminant could be swamped by protein, on a weight basis, in a preparative-scale run where the protein content of a peak is high and where pre-running has brought the 'background' to A_{280} of below, say, 0.2 (and, hopefully, reduced A_{210} also). Yet the analysis of amino acids in hydrolysates of separated proteins may be vitiated through the appearance of rogue ninhydrin colours and even of a white precipitate (S. Jacobs, personal

Table 1. Typical finding for UV aborption and ninhydrin reactivity of the contaminant 'X' (as obtained by leaching gels), and its elementary composition.

Feature	Observed values for 'X'	Observed value for acrylamide	Observed values for other materials tested
Dry wt. corresponding to 1 'density unit' at 210 nm (i.e. A_{210} = unity if present in 1 ml, with 1 cm light path)	50-100 μg	11 μg	Bovine serum albumin, 58 μg
A_{280}/A_{210} ratio	<0.01-0.3 but often 0.03 -0.1. (cf. value <0.05 calc. from ref. 4) (A_{260}/A_{210} similar)	0.01	Acrylic acid similar to acrylamide (peak at 210 nm); bovine serum albumin, 0.04
Ratio of A_{570} in ninhydrin reaction (pH 5) to A_{210} (with equivalent volumes)	Usually 0.5-2.5 (but may be <0.4 with a hot polymerized gel)	0.06 (but may be of the same order as the 'X' value on a wt. basis)	Polyacrylamide, negligible; TEMED (tetramethylethylene diamine), 0.035; DMAP (3-dimethylaminopropionitrile), 0.015
Ninhydrin reaction (pH 5): ratio of A_{400} to A_{570}	Usually 0.1-0.3		Lysine*, 0.4; ammonia, 0.37 - but 0.1 reported (13) for ammonia (or for leucine; 2.0 for proline)
Elementary composition, % (including N liberated as NH_3 by refluxing for 30 min in 1.25N NaOH)	C, 48.6; H, 8.4; N, 21.4; (21.3†)	*Calc.:* C, 50.7; H, 8.4; N, 19.7 (20.3†)	*Calc. for 19:1 acrylamide -'BIS' mixture:* C, 50.8; H, 7.05; N, 19.6** (Polyacrylamide, 8.4†)

* Weight-for-weight, lysine gave 3 times the A_{570} colour typically found with 'X'.

† Observed value, for N liberated as NH_3 by NaOH hydrolysis

** This calculated value takes no account of any traces of accelerator, for which the % N is high ('TEMED', 24.1; 'DMAP', 28.5)

communication). The ninhydrin-reactivity of 'X', as also encountered in another laboratory (7), has shown 5-fold variations (Table 1), uncorrelated with the variations in the UV spectrum. A discouraging further observation (not tabulated) is that with 6N HCl treatment of 'X', simulating protein hydrolysis for amino acid analysis, the ninhydrin A_{570} may rise 8-fold; the 'lysine equivalent' of 1 µg of 'X' typically rises from 0.3 to 2 µg (reflecting liberation of ammonia?).

The possibility was explored that the ninhydrin colour from 'X' might be reduced relative to that of amino acids by trying variations (13) in the conventional procedure of heating at pH 5.0 and reading at 570 nm. As is illustrated by the ninhydrin A_{400}:A_{570} values given in Table 1 for 'X' (note the variability) and for certain amino acids, the hope of finding a clear-cut difference in absorption spectra was not realized. Changing the ninhydrin reaction pH likewise gave unhelpful results (not tabulated).

Elementary analysis and general comparison with acrylamide

Table 1 shows that the elementary composition of 'X' is akin to, if not identical with, that of acrylamide. All the nitrogen appears to be amide-N as judged by NH_3 estimation (by distillation) after alkaline hydrolysis; this procedure gives only 43% of the theoretical value if applied to polyacrylamide, in accordance with a report (14) that the rate of chemical hydrolysis of polyacrylamide slows markedly once 40-50% has been hydrolyzed.

To summarize, 'X' is acrylamide-like in elementary composition, whilst differing from acrylamide in 210 nm absorption and ninhydrin colour yield per unit dry weight. It differs in apparent 'amide-N' content, and in ninhydrin reactivity, from polyacrylamide gel, but might well consist of an 'oligo-acrylamide' that yields NH_3 on alkaline hydrolysis as readily as acrylamide itself.

FEATURES OF THE CONTAMINANT RELEVANT TO 'DECONTAMINATION'

Insofar as 'X' may be an undesirable contaminant of the macromolecular separation products in a gel run, and would vitiate determinations particularly if based on ninhydrin colour, a reliable method for its removal is needed. In the following studies with gel leaches, the amount of 'X' was assessed by A_{210} readings, with prior perchloric acid treatment where protein could interfere.

Selective precipitation

Most proteins are precipitable by perchloric acid addition (to 5%) in the cold, whereas 'X' is not thus precipitated if tested in the absence of protein. In principle, then, protein present in admixture with 'X' could be estimated by difference, e.g. from A_{280} readings taken before and after perchloric acid treatment; 'X' could be determined directly in the supernatant. In practice there will be an error because, as it turns out, there is some co-precipitation of 'X' with protein when a protein precipitant is added to such a mixture.

Non-denaturing precipitation procedures would obviously be very advantageous. 'X' is soluble at all pH values; but too few proteins furnish isoelectric precipitates for this approach to be widely applicable, even if co-precipitation proved to be minimal. Precipitation of separated proteins (or of RNA) by cold ethanol does not represent a useful general approach, since 'X' when tested alone is precipitable by ethanol even in the absence of protein, but it might work with certain proteins if sharply different from 'X' in ethanol-precipitability. In principle it might be feasible to separate out 'X' preferentially by a salting-out procedure, particularly in view of the possible effectiveness of electrolytes at fairly high concentration in precipitating lyophilic sols. Whilst 'X' may well be such a sol, the salting-out approach was not felt to be so widely applicable as to warrant a pilot study.

Methods entailing use of a membrane

The contaminant has been variously described as "nondialyzable" (2) and ultrafiltrable (8). In our experience it is partly dialyzable. It is partly ultrafiltrable with a membrane which retains typical protein (Amicon, nominal 'cut-off' at 10,000). However, its size distribution overlaps with that of proteins as judged by filtration experiments at the National Institute for Medical Research (S. Jacobs, personal communication).

Electrodialysis has also been tried, over the pH range 3-11, with Visking or Amicon membranes bounding the central compartment into which the buffered leach was put initially. After the electrodialytic run (about 2 h) the contents of all three compartments showed A_{210} readings and ninhydrin colours; indeed, the recovery of A_{210} units was often more than than 200%, as had also been found in the ultrafiltration experiments. The material that gave rise to these readings seemed to be largely cationic even at pH 11 (cf. 7), although some of it moves to the anode whatever the pH. It is well known that lyophilic sol particles are heterogeneous in charge distribution as found with the present material.

Whilst 'X' evidently can in part pass through a membrane when tested in the absence of protein, in the presence of protein its passage is impaired, as shown by subjecting mixtures of 'X' and bovine serum albumin to conventional dialysis, ultrafiltration, or electrodialysis.

Molecular-sieve chromatography

With G-150 Sephadex, Charlwood (5) reckoned that he had removed from human β_{1A} globulins, as separated by polyacrylamide gel electrophoresis and finally examined in the analytical ultracentrifuge, a "small amount of slowly sedimenting substance that derived from the polyacrylamide gel". However, our own chromatographic work as summarized below has not been encouraging.

As model proteins we chose cytochrome *c* and bovine serum albumin (both from BDH Chemicals Ltd.), for which the molecular weights are reckoned to be about 12,000 and 67,000 respectively. These proteins are readily available, but cytochrome *c* has the disadvantage that perchloric acid does not precipitate it from weak solutions in the absence of other proteins; bovine serum albumin has the disadvantage that, as we indeed found, multiple peaks may appear, due to molecular association (15). Where protein and 'X' were to be run conjointly, these were pre-mixed. The 'X' was leached material of low A_{280}.

The column used for each set of runs contained a Sephadex bed of height 270 mm (or 150 mm in the case of G-150) and of diameter 25 mm; it was run at room temperature. The runs were at pH 8.5 in 12.5 mM borate/ 15 mM HCl, this buffer having a conveniently low A_{210} value. The effluents were examined at 410 nm (for cytochrome) 280 nm (for albumin, and for cytochrome so as to confirm its presence), and 210 nm (for 'X'). For interpreting the A_{210} values, account had to be taken of the contribution due to any protein that was present in the fraction. Persistence of some absorption at 210 nm even after deproteinizing with perchloric acid (to 5%) served to indicate that some 'X' was present. The finding of a positive ninhydrin colour (570 nm) after deproteinization was likewise indicative of 'X'; but in view of the above-mentioned variability of the ninhydrin colour, the absence of such colour could not be held to prove the absence of 'X'. In view of the lack of a specific assay for 'X', and of apparent over-recovery of 'X' in the fractions as judged by A_{210} readings, the results for 'X' were essentially qualitative than quantitative.

With G-100 Sephadex, rather similar behaviour (elution largely in the range 80-120 ml) was observed for the different test materials, including 'X', when run individually. Whilst it was encouraging that 'X' did run largely in one region (90-125 ml, as a double peak), although it also 'smeared', protein and 'X' loaded in admixture emerged together. Only the late part of the cytochrome peak, which seemed to be retarded when co-chromatographed with 'X', was low in 'X'. On G-150, 'X' showed peaks at 130-155 ml and at 170-210 ml, and albumin at 50-60 ml and 110-120 ml. Albumin emerged earlier together with 'X' if run admixed with 'X'; only a non-retained albumin peak was free from 'X'. G-200 gave a similar result, the positions of albumin and 'X' being earlier and partly coincident when a mixture was chromatographed. The cytochrome *c* position on G-150 or G-200 was too similar to that of 'X' to be propitious for attempting separation from 'X'.

In summary, the studies with Sephadex did show that 'X' in the absence of protein may run largely as a definite peak or peaks, not solely as a 'smear'. However, there was rather little difference in position from the model proteins studied and their chromatographic behaviour was altered when 'X' was present. 'X' seems to have an avidity for proteins such that a mixture of 'X' and a protein was not well resolved under the conditions used. We have not explored the possibility that separation might be achievable by working at pH more acid than the isoelectric point of the protein concerned, possibly with carboxymethylcellulose (alone, or as CM-Sephadex) rather than with Sephadex as now used. Whilst Charlwood (5) refers to the efficacy of Sephadex as mentioned above, he does not state whether his gel impurity (presumably 'X') was detectable in his chromatographic eluate in a position different from that of the protein. Chrambach and Rodbard state in their review (2), without documentation, that the contaminant can be separated from proteins by use of G-50 Sephadex. Separation by Sephadex may well be more successful with large macromolecules such as globulins than with albumins.

INERTNESS OF THE IMPURITY 'X' IN CERTAIN TEST SYSTEMS

Winters *et al.* (6) stated that the material which was inadvertently eluted along with the desired adenovirus antigens had "no effect on immunological properties or biological activities".

With an apparatus for preparative gel electrophoresis (see *Article 2* in this book, by A.D. Brownstone, who alludes to the contamination problem), we have prepared an 'X'-containing eluate taken 9 h after the start of the run, to examine its possible inhibitory action towards alcohol dehydrogenase. The enzyme (Sigma Chemical Co.) was assayed at 25° in an NAD-containing medium (10) to which one volume of the eluate was added. Since dehydrogenases are quite sensitive to poisoning, it is reassuring that no inhibition was observed, even if the enzyme were kept in contact with 'X' for 3 h at 2° before the incubation.

CONCLUDING REMARKS

We have confirmed and examined a phenomenon already mentioned briefly by other workers who have used polyacrylamide for preparative purposes, viz. the appearance in the eluates of a material (here termed 'X') which is derived from the gel itself, and for which an S value of 1.5 has been reported (1). 'X' has its absorption peak at or just below 210 nm although the peak position is variable, but may also show sufficient absorption at 280 or 260 nm to confuse the detection and analysis of separated macromolecules, and indeed may give a strong ninhydrin colour, particularly after 6N HCl hydrolysis. Presumably it was the presence of this contaminant *in situ* that precluded UV scanning of slab gels unless the materials had been recrystallized at the outset (9) - a precaution which has proved ineffective in our hands. The contaminant awaits study in relation to colour reactions for nucleic acid constituents, but at least does not interfere with examination of separated RNA bands in electrophoretic gel slabs subjected to staining (G.N. Dessev, personal communication). 'X' admixed with protein remains partly in solution when perchloric acid is added, and can be estimated by reading the supernatant at 210 nm.

There is circumstantial evidence that the material consists largely of non-latticed polyacrylamide ('oligo-acrylamide'?) perhaps formed continuously during electrophoresis or leaching tests, which differs from the latticed polymer in giving a quantitative yield of NH_3 in an amide-N determination and in giving, albeit erratically, a definite colour in the ninhydrin test. (It can hardly be literally "linear polyacrylate" (3).) In general the material behaves like a lyophilic sol, with a heterogeneous charge distribution and variable optical behaviour in the UV region; it appears to revert to a gel on drying down. Possibly it is formed within the gel by a gel-sol transition which continues almost indefinitely, as would seem to be the case in leaching experiments. An alternative possibility is that non-latticed polymer molecules are formed only during the initial polymerisation and escape only slowly because of mechanical retardation in the gel lattice. Positive evidence for the latter possibility might be obtainable by trial of a gel column with a thin layer of labelled gel as its base; with electrophoresis the radioactivity of the eluate should fall off more quickly than the A_{210} readings.

Some reduction in the A_{280} baseline due to 'X' may be achieved if the precaution is taken of doing an electrophoretic pre-run before loading the sample - say for 5 h or, better, overnight; with longer pre-running the gel might deteriorate prematurely. Despite a fall in the A_{280} baseline (cf. Fig. 2 in ref. 1) the A_{210} baseline may remain high, and ninhydrin peaks may appear several hours after the start of the pre-run. Our leaching experiments hardly vindicate the effectiveness of soaking the gel block before putting it into the apparatus; but this was felt to be a useful precaution in transferrin studies (D. Ramsden & A. Tavill, personal communication), and a strong salt solution, e.g. 0.5 M ammonium sulphate, has been recommended with a final wash-out (C.J.O.R. Morris).

Evidently, then, one must learn to 'live with' this material, the formation of which seems to be inevitable. A particular difficulty is that it tends to associate with proteins. There is no general method for removing it. Despite our discouraging results, a chromatographic step might be effective with some samples, and is in any case desirable for the sake of checking and possibly improving the extent to which the macromolecular bioconstituents have been separated from one another. Removal of the contaminant is hardly likely to be achieved by conventional dialysis, electrodialysis or membrane filtration. Fortunately, the contaminant may be innocuous in 'biological' systems, as instanced by alcohol dehydrogenase.

Acknowledgements

Mrs. Ann C. Cunningham and Mr. A.R. Jones assisted in the work, and Dr.J.P. Leppard made valuable comments. Shandon Southern Insts. Ltd. gave support.

Mr. A. Brownstone and colleagues kindly commented on a draft.

References

1. Hjertén, S., *J. Chromatog. 11* (1963) 66.
2. Chrambach, A. & Rodbard, D., *Science (Wash.) 172* (1971) 440.
3. Jovin, J., Chrambach, A. & Naughton, M.G., *Anal. Biochem. 9* (1964) 351.
4. Gordon, A.H., in *Laboratory Techniques in Biochemistry and Molecular Biology* (T.S. Work & E. Work, eds.), North Holland, Amsterdam (1969), p. 1.
5. Charlwood, P.A., *Biochem. J. 115* (1969) 897.
6. Winters, W.D., Brownstone, A. & Pereira, H.G., *J. Gen. Virol. 9* (1970) 105.
7. Van Kreel, B.K., Pijnenburg, G.M.C.M., Van Eijk, H.G. & Leijne, B., *Clin. Chim Acta 32* (1971) 103.
8. Kaltschmidt, E. & Wittmann, H.G., *Anal. Biochem. 30* (1969) 132.
9. Loening, U.E., *Biochem. J. 102* (1967) 251.
10. Riggs, J.P. & Rodriguez, F., *J. Polymer Sci. 1A, 5* (1967) 315.
11. Gordon, A.H. & Lewis, I.N., *Anal. Biochem. 21* (1967) 190.
12. Bailey, J.L., *Techniques in Protein Chemistry,* Elsevier, Amsterdam (1962).
13. Moore, S. & Stein, W.H., *J. Biol. Chem. 176* (1948) 367.
14. Smets, G.J., in *Chemical Reactions of Polymers* (E.M. Fettes, ed.), Interscience, New York (1964).
15. Anderson, L.-A., *Biochim. Biophys. Acta 117* (1966) 115.
16. Hinton, R.H., Burge, M.L.E. & Hartman, G.C., *Anal. Biochem. 29* (1969) 248.

4 PREPARATIVE GEL ELECTROPHORESIS: RECOVERY OF PRODUCTS *with a Note on* Free Zone Electrophoresis

Stellan Hjerten
Institute of Biochemistry
University of Uppsala, Box 531,
3-751, Uppsala 1, Sweden

Elution chambers of various types are commonly used in preparative gel electrophoresis. Two novel recovery techniques are described, both involving cutting the gel into slices after completed electrophoresis. In one of these methods the substance to be recovered is allowed to migrate electrophoretically out of the gel slice into a sucrose gradient, where the substance is concentrated to a narrow band; the other method is based upon a dissolution of the gel slice. These two methods are particularly valuable for the separation of small amounts of material.

Analytical molecular-sieve electrophoresis, for instance in polyacrylamide gels, is characterized by an extremely high resolving power. When the method is transferred to a preparative scale the practical resolving power is seriously diminished unless considerable attention is paid to the manner in which the zones are recovered from the gel. Therefore in the following discussion of preparative gel electrophoresis, particular emphasis is placed on the recovery techniques.

A. RECOVERY BY A CONTINUOUS BUFFER FLOW FROM ELECTROPHORESIS

In this technique the substances are transferred by a buffer flow to a fraction collector when they emerge electrophoretically from the gel column. In a previous paper we have compared the different principles that have been utilized for this elution technique (ref. 1, particularly pp. 123-125). We found that the recovery process was most efficient if the elution chamber contained a supporting medium such as Sepharose 4B (Fig. 1). Two preparative gel electrophoresis columns have been constructed for this elution method, one for about 5-50 mg of sample material[x] (2) and the other with a maximum capacity of around 1 gram[†] (1). The resolving power of these columns is equivalent to that obtained on an analytical scale.

All methods utilising a continuous elution technique give a dilution of the sample zones, which therefore may be difficult to detect

x *Commercially available from Stålprodukter, PO Box 12036, S-750 12 Uppsala, Sweden.*

† *Will be commercially available from LKB-Produkter AB, S-161 25 Bromma 1, Sweden*

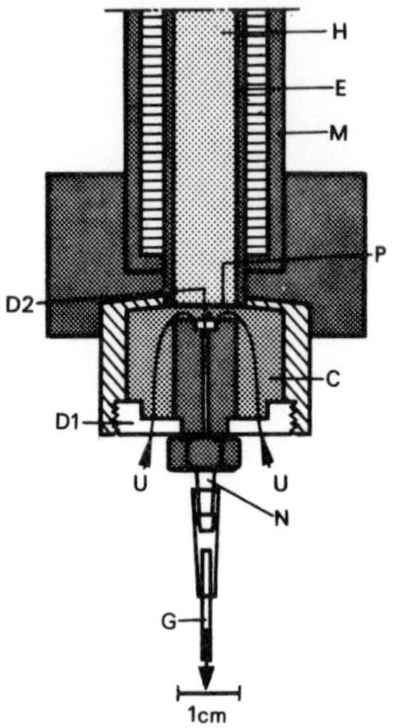

Fig. 1. The principle of recovery by a continuous buffer flow during electrophoresis. *E = electrophoresis tube; H = polyacrylamide gel column; M = cooling mantle; C = elution chamber, filled with Sepharose 4B; P = filter paper; D1,D2 = porous polyethylene disks; U = buffer flow transferring substances to a fraction collector as they leave electrophoretically the gel column, H; G = polyethylene tubing leading to a fraction collector; N = nipple.*

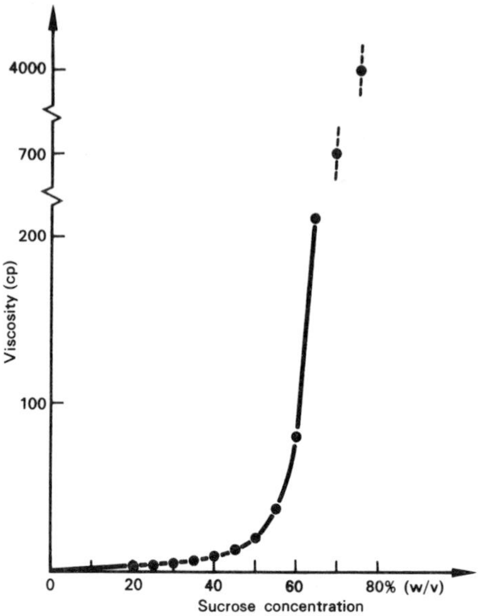

Fig. 2. The viscosity of a sucrose solution as a function of the sucrose concentration (4).

in the eluate when the amount of material present is lower than 5-10 mg. In such cases the recovery techniques B and C described below are more suitable.

B. RECOVERY OF THE SAMPLE ZONES BY SEGMENTATION OF THE GEL FOLLOWED BY ELECTROPHORESIS OF THE INDIVIDUAL SLICES AGAINST A SUCROSE-NaCl GRADIENT

In Fig. 2 the viscosity of a sucrose solution is plotted against the sucrose concentration. The viscosity increases rapidly when the sucrose concentration exceeds 40% (w/v). Since the electrophoretic migration velocity decreases when the viscosity is increased, buffer salts and applied sample substances will accummulate during electrophoresis in a sucrose gradient, provided the sucrose concentration is sufficiently high. The attendant increase in the electrical conductivity will further decrease the migration velocity of the sample substances, thereby enhancing the efficiency of the concentration; this zone-sharpening effect can be strengthened artificially by including salts, such as sodium chloride, in the sucrose solutions. A more detailed treatment of this concentration method is to be found in a paper (3) where some other factors that contribute to decreasing the migration velocities in the sucrose gradient are also discussed.

This accumulation of sample substances in a sucrose gradient when an electric field is applied has been utilized not only for concentration purposes, but also for the removal of neutral polymers used in partition and precipitation experiments (5), and for the recovery of substances following a polyacrylamide gel electrophoresis (3). The latter procedure will now be described in detail.

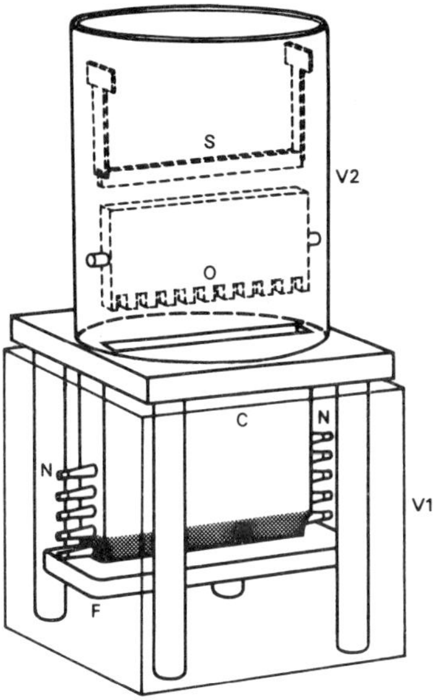

Fig. 3. The electrophoresis apparatus. *Material: perspex. C = electrophoresis chamber (inner dimensions: 0.5 x 10 x 10 cm); V1,V2 = electrode vessels; M = dialysis membrane; F = frame to attach the membrane M to the electrophoresis chamber C; O = mould; which is inserted into electrophoresis chamber C when casting gels for the application of up to nine different samples; S = gel slice support inserted into the chamber C and used for electrophoretic recovery of a substance from a gel slice; N = nipples (made of PVC).*

Apparatus and methods

*The construction of the apparatus** can be varied according to the demands of the particular problem at hand. The version depicted in Fig. 3 has been used in the author's laboratory not only for the recovery technique described herein but also for conventional analytical polyacrylamide gel electrophoresis, especially when comparative analyses of many different samples are to be performed.

* *commercially available from Stalprodukter, P.O. Box 12036, S-750 12 Uppsala, Sweden*

Casting the polyacrylamide gel. The dialysis membrane M, moistened with buffer, is firmly attached to the bottom of the electrophoresis chamber C with the aid of the frame F. The electrode vessel V1 is filled with buffer almost to the brim, so that most of the electrophoresis chamber is surrounded by buffer. This water-cooling of the electrophoresis chamber provides effective dissipation of the heat generated during the electrophoresis. The cooling efficiency is further improved if the buffer is stirred, for instance by a magnetic stirrer. The deaerated monomer solution, containing initiator and accelerator, is poured into the electrophoresis chamber, and mould O inserted. (For preparative runs we have not employed the mould but have created an even upper gel surface by layering buffer on top of the monomer solution.) When the polymerization is complete the mould is removed. The upper electrode vessel V2 is then filled with buffer.

The electrophoresis step. Following pre-electrophoresis for about 30 min to remove the initiator (persulphate) and some UV-absorbing material, the sample(s) can be applied by layering. If it is necessary to increase the density of the sample by addition of sucrose to facilitate the application, only small amounts of sucrose should be used, lest the electrophoretic zones be distorted. After completion of the run the membrane M is removed and the gel is pressed out by inserting a perspex sheet into the gel chamber.

Localization of the zones in the gel. For preparative runs a 5 mm wide strip or, preferably, two or three at appropriate distances from each other, are cut out of the gel slab in a direction parallel to that of the electrophoretic migration. By staining (or UV-scanning) one can locate the zones in the strips - and consequently in the major part of the gel slab. The zone of interest is then cut out of the main gel slab and placed on the support S inserted into the electrophoresis chamber C, which has previously been filled with a sucrose-NaCl gradient as described in the next section. The same apparatus (Fig. 3) can be used for the gel electrophoresis and for the subsequent recovery of the substance from the gel slice.

The sucrose-NaCl gradient. To increase the conductivity and hence the concentrating effect, sodium chloride (or another inorganic salt) is in general included in the sucrose gradient. The composition and the method used for the preparation of the gradient are not critical. A typical example is the procedure used in the experiment corresponding to Fig. 5 below. (Another gradient has been described elsewhere (3)). In this case six different solutions (S_1-S_6) were prepared, the compositions of which are given in Table 1. By means of an inverted pipette (the viscosity in solution S_1 is very high!) the chamber C is filled with solution S_1 up to a height of 3-4 cm. On top of S_1 a 2-mm layer of solution S_2 is pipetted and on S_2 a 2-mm layer of solution S_3, and so on.

The dialysis membrane M should be surrounded by solution S_1 during the electrophoresis. To spare sucrose and yet fulfil this requirement a smaller vessel B, containing S_1, is placed in electrode vessel V1 (see Fig. 1 in ref. 3). After some hours the sucrose-NaCl gradient is sufficiently smooth to stabilize against convection, The gel slice containing the zone of interest is placed in the electrophoresis chamber on the support S (Fig. 5a). Buffer without sucrose and NaCl is then layered on top of the gradient.

Table 1. Compositions of the solutions used for the preparation of the sucrose-NaCl gradient. *The buffer concentration is about the same in all the solutions; namely 0.22 M glycine-NaOH (pH 8.8).*

	Sucrose concn., % (w/v)	NaCl concn., % (w/v)
S_1	75 (= 2.2 M)	12 (=2.0 M)
S_2	60	9.6
S_3	45	7.2
S_4	30	4.8
S_5	15	2.4
S_6	7.5	1.2

Emptying the column. After electrophoresis for a time determined by the migration velocities of the sample substances, the zone has migrated out of the gel slice and has become concentrated in a narrow zone in the sucrose gradient (Fig. 5b). There is very little risk that the zone will leave the gradient even if the concentration is allowed to proceed overnight (Fig. 5c). The electrophoresis chamber is then emptied by connecting a syringe to the upper one of the nipples N. After the sucrose solution above this nipple has been slowly withdrawn, the syringe is loosened and connected to the next nipple, and so on. (Alternatively one can empty the column by withdrawing 1-cm fractions from the top.) By measuring the UV-absorption of the different fractions obtained - or, when the concentrations of the sample substances are low, by measuring their UV-spectra - the recovered substance can be localized in the sucrose gradient (Fig. 6). It should be pointed out that the solute zones will not, in general, reach regions in the gradient where the sucrose or NaCl concentration is extremely high (see Fig. 5b). When necessary, the sucrose and NaCl can be removed from the sample by dialysis, ultrafiltration, ion-exchange chromatography, or molecular sieving on a tight gel such as Sephadex G-25 or Bio-Gel P-10. The recovered substances seem to be contaminated by some gel material having a sedimentation coefficient of about 1.5 S (3).

In order to illustrate the recovery technique, an artificial mixture of two coloured substances (naphthol green and haemoglobin) was used. After separation of these substances (Fig. 4) in the polyacrylamide gel electrophoresis apparatus depicted in Fig. 3, the gel was removed and cut into slices perpendicular to the direction of the electrophoretic migration. The same apparatus was then filled with a sucrose-NaCl gradient. The gel slice G containing the haemoglobin was placed above the gradient R

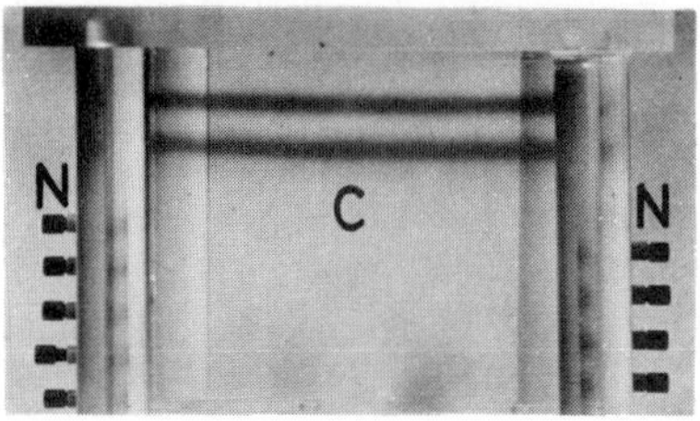

Fig. 4. Separation of an artificial mixture of haemoglobin and naphthol green by polyacrylamide gel electrophoresis in the equipment depicted in Fig. 3.

(Fig. 5a). After 70 min of electrophoresis the haemoglobin had left the gel slice and sharpened into a narrow zone in the sucrose gradient (Fig. 5b). To show that prolonged electrophoresis does not diminish the concentration effect, the run was allowed to proceed overnight (Fig. 5c). The column was then emptied and the haemoglobin concentrations in the different fractions were determined by absorption measurements (Fig. 6).

C. RECOVERY OF THE SAMPLE ZONES BY SEGMENTATION OF THE GEL FOLLOWED BY ENZYMATIC DEGRADATION OF THE INDIVIDUAL SLICES

Unfortunately, no enzyme capable of degrading a polyacrylamide gel has yet been discovered. Therefore we have investigated the use of dextran gels as supporting and sieving medium for electrophoresis, since these gels can be degraded easily by dextranase. The combination agar(ose)-agarase can also be utilized but will not be treated herein, partly because there exists already a simple method for the removal of agarose after electrophoresis (spinning down of the agarose from a 0.15-0.17% agarose suspension (6)).

Preparation of the dextran gel. These gels have been prepared essentially as described by Flodin (7), using epichlorohydrin as the cross-linker. The gels have had about the same dimensions as the polyacrylamide gels mentioned in the preceding sections (0.3-0.5 x 10 x 10 cm). Since the cross-linking with epichlorohydrin requires a high pH, the gel must be detached from the mould and equilibrated with the buffer to be used for the subsequent electrophoresis - a procedure which also removes all diffusible products formed during the polymerization. During this washing the gels swell considerably.

The run. The dextran gel slabs have been inserted into a modified version of the apparatus depicted in Fig. 3.* The run and localization of the zones are performed as described previously for the vertical polyacrylamide gels. (Since the dextran gels have the great advantage of very low UV-absorption the zones can be detected easily by direct scanning.) The gel slice containing the substance of interest is cut into pieces and transferred to a test tube to which a small volume of a dextranase solution is added. Upon incubation overnight the dextran gel is solubilized. The dextranase was prepared according to the method described in ref. 8. The enzyme concentra-

* *The dextran gels will probably be commercially available from Pharmacia Fine Chemicals AB, Uppsala, Sweden.*

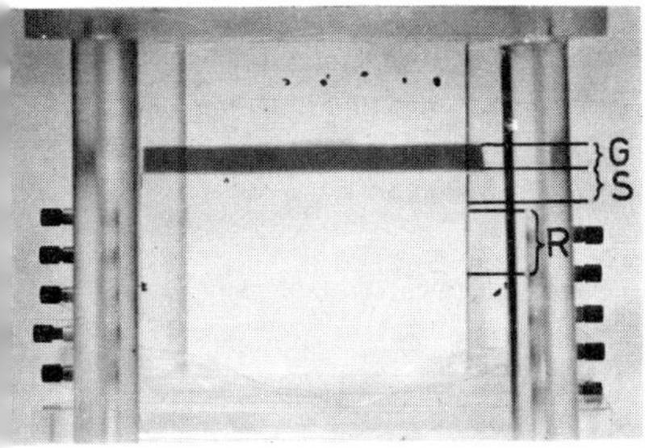

5a

5b

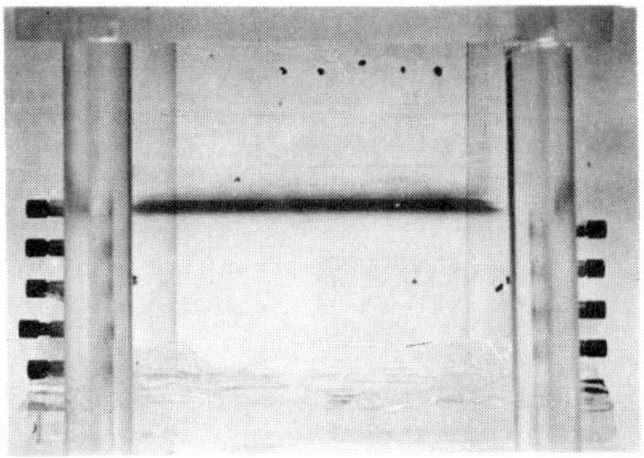

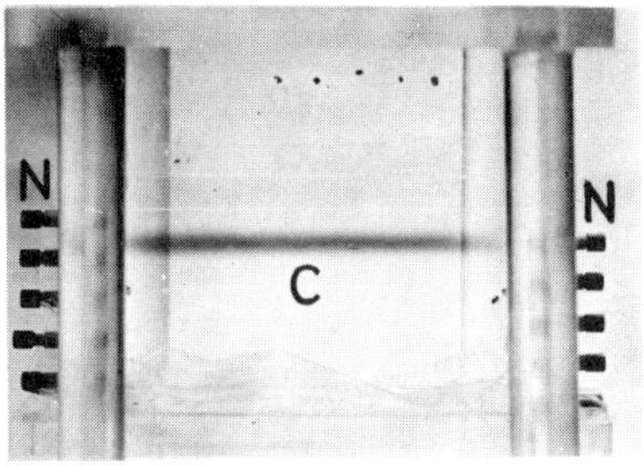

5c

Fig. 5. Recovery of a haemoglobin following the gel electrophoresis photographed in Fig. 4.

a) *The gel slice containing the haemoglobin rests on a support S (shown in detail in Fig. 3) above a sucrose-NaCl gradient R.*

b) *Photograph taken after 70 min of electrophoresis at 37 V (30 mA). The haemoglobin has migrated out of the gel slice and is concentrated in a narrow zone in the sucrose-NaCl gradient.*

c) *After electrophoresis for an additional 16 h at 10 mA the zone is still sharp.*

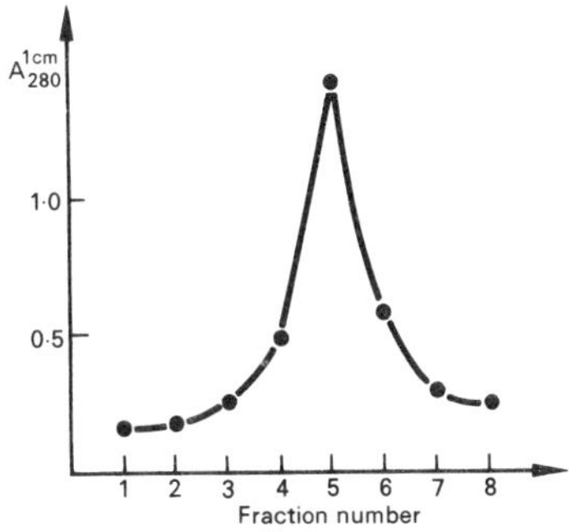

Fig. 6. Localization in the sucrose-NaCl gradient of the recovered haemoglobin. *After emptying of the electrophoresis chamber C, shown in Fig. 5c, by withdrawing the gradient successively through the nipples N, the absorbancy at 280 nm of each of the eight fractions was determined.*

tion required for solubilization of the dextran gel was so low that the enzyme was difficult to detect by absorption measurements at 280 nm in the solution obtained after liquification of the gel. The contamination arising from the addition of dextranase, which has a molecular weight of 60,000, is therefore often negligible. The dextran is broken down so completely that most of the degradation products can be removed by dialysis.

Illustrative experiments

The author has previously pointed out that dextran gels should exhibit molecular-sieving properties in electrophoresis as they do in chromatography (9). This assumption is confirmed by the experiments shown in Figs. 7 and 8.

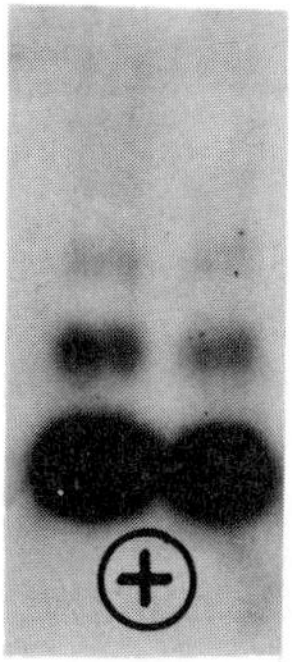

Fig. 7. Molecular-sieve electrophoresis in a dextran gel. *Sample: human serum albumin (monomer, dimer, trimer, and tetramer). Buffer: 0.1 M Tris-HCl, pH 8.0. Voltage: 22V. Current: 40 mA. Time: 22 h. Staining agent: Amido Black.*

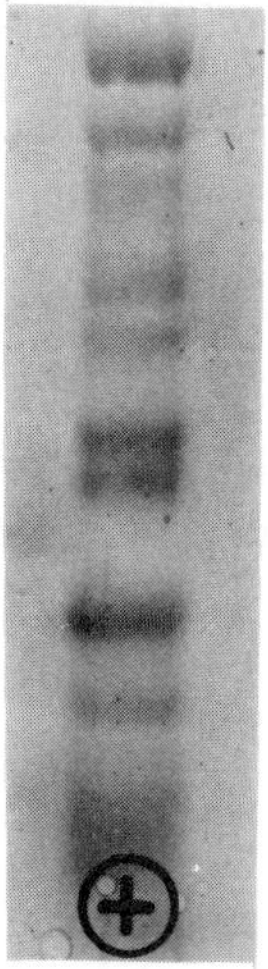

Fig. 8. Molecular-sieve electrophoresis in a dextran gel. *Sample: membrane proteins from* Mycoplasma laidlawii. *Buffer: 0.1 M Tris-HCl, pH 8.0 + 0.02 M sodium dodecyl sulphate. Voltage: 40 V. Current: 75 mA. Time: 8 h. Figs. 7 and 8 illustrate the exquisite molecular sieving properties of dextran gels (in non-sieving media, such as agarose gels, no separation was observed).*

It is well known that gels of starch, polyacrylamide, and agarose (in sufficiently high concentration) also exhibit molecular sieving in both electrophoresis and chromatography (see refs. 9 and 10). Therefore - and also for other reasons (11) - the author prefers the analogous terms molecular-sieve electrophoresis and molecular-sieve chromatography in these and similar gels.

Acknowledgements

The author is much indebted to Mrs. Irja Blomqvist for the performance of all the experiments described, and to Dr. Jan-Christer Janson for the dextranase preparation.

The work has been supported by grants from the von Kantzow Foundation, the Swedish Natural Science Research Council, and the Wallenberg Foundation.

References

1. Hjertén, S., Jerstedt, S. and Tiselius, A., *Anal. Biochem. 27* (1969) 108.
2. Hjertén, S., Jerstedt, S. and Tiselius, A., *Anal. Biochem. 11* (1965) 211.
3. Hjertén, S., *Biochim. Biophys. Acta 237* (1971) 395.
4. Landolt-Bornstein, *Zahlenwerte und Funktionen aus Physik, Chemie, Astronomie, Geophysik, Technik, Transportphänomene I, Vol. 2 Part 5*, Springer Verlag (1969) pp. 377 & 477.
5. Albertsson, P.-Å., *Adv. Prot. Chem. 24* (1970) 309.
6. Hjertén, S., *J. Chromatog. 12* (1963) 510.
7. Flodin, P., Thesis, Meijels Bokindustri, Halmstad, Sweden 1962.
8. Janson, J.-C. and Porath, J., *Methods in Enzymol. 8* (1966) 615.
9. Hjertén, S., in *Protides of the Biological Fluids* (H. Peeters, ed., Proceedings of the 14th Colloquium, Bruges, 1966), Elsevier, Amsterdam (1967) p.533.
10. Morris, C.J.O.R. & Morris, P., *Biochem. J. 124* (1971) 517.
11. Hjertén, S., in *New Techniques in Amino Acid, Peptide and Protein Analysis* (A. Niederwiser & G. Pataki, eds.), Ann Arbor, Michigan, U.S.A. (1971) p. 227.

A Note on FREE ZONE ELECTROPHORESIS

Free zone electrophoresis is a new, rapid electrophoresis method which is intended primarily for analytical purposes, but which is very useful also for preparative experiments on a small scale (as is described in detail elsewhere) (17). The amount of material required is about the same as in paper electrophoresis (10 µl, 3-50 µg). The maximum capacity is about 400 µg. Free zone electrophoresis is characterized by a very broad range of applications. Thus it has been used for investigation of low-molecular inorganic and organic ions, amino acids, peptides, proteins, nucleic acid bases, nucleosides, nucleotides, nucleic acids, viruses, ribosomes, erythrocytes and bacteria. An example of a protein separation is given in Fig. 9 (2). The apparatus* can be utilized not only for conventional electrophoresis but also for isoelectric focusing and displacement electrophoresis.

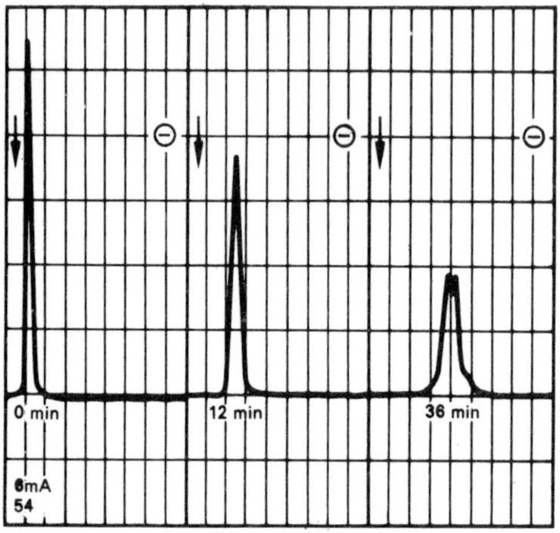

Fig. 9. Free zone electrophoresis of a rape seed protein fraction. *Buffer: 0.05 M Tris-HCl, pH 8.0. Inner diameter of the revolving electrophoresis tube: 3mm. Current: 6 mA. Voltage: 1100 V. Sample amount: 20 µg in 7 µl of buffer. The scans were made at the times indicated. The arrow indicates the position of the starting zone.*

The separation chamber is a horizontal quartz tube which slowly rotates around its long axis (40 rev/min). The rotation eliminates the need for a supporting medium, the presence of which in many cases can be very disturbing and sometimes make an analysis impossible. The free zone electrophoresis apparatus is fitted with a UV-scanning system with the aid of which the different zones in the quartz tube are located during a run. A zone in the tube corresponds to a peak on the diagram of the recorder. It should be pointed out that even substances which have no UV-absorption can be detected if a suitable buffer is used.

The free zone electrophoresis equipment is fully automated, i.e. after the application of the sample no further attention is required.

References (Free Zone Electrophoresis)

1. Hjertén, S., *Chromatog. Rev. 9* (1967) 122.
2. Lönnerdal, B. and Janson, J.-C., *Biochim. Biophys. Acta 278* (1972) 175.

* *Commercially available from Incentive Research & Development AB, P.O. Box 11074, S-161 30 Bromma 11, Sweden.*

5

A SIMPLE PREPARATIVE METHOD FOR SEPARATION OF RNA AND RIBOSOMAL SUBPARTICLES BY ELECTROPHORESIS IN AGAR AND POLYACRYLAMIDE GELS

G.N. Dessev and K. Grancharov
Institute of Biochemistry
Bulgarian Academy of Sciences
13 Sofia
Bulgaria

In the procedure now described, pre-'staining' with acridine orange enables separation and elution to be watched whilst in progress.

The present technique has been developed on the basis of previously described analytical procedures (1,2). It takes advantage of the fact that RNA and ribosomal subparticles adsorb acridine orange (3), with little effect of the electrophoretic behaviour of the fractions. In this way it is possible to follow visually the course of the separation and elution.

A simple set-up for column gel electrophoresis has been constructed (Fig. 1), but any other apparatus for preparative fractionation could be employed. The gel (1-25% agar or 2.5-2.8% polyacrylamide, each in 0.025 M Tris-acetate, pH 7.6) is cast in a glass tube closed with a dialysis membrane as shown in Fig. 1. The same buffer circulates through the system, washing the electrodes.

The method has been tested with rat liver ribosomal RNA isolated by SDS and phenol, and with ribosomal subparticles obtained by EDTA dissociation of rat liver polysomes (4). The sample containing in both cases about 30 A_{260}-units/ml, is mixed with acridine orange (0.02-0.04 ml of a 1% solution of the dye per ml of the sample) and sucrose is added to a final concentration of 6%. A beautiful yellow-green fluorescence develops, which is best observed against a black screen. Traces of detergents might quench the fluorescence but the addition of 0.1 vol of ethanol neutralized this effect without affecting the separation. The column is filled with buffer and the sample is underlayered. The amount of material should not exceed 7-8 A_{260}-units/cm^2. The electrophoresis is performed at 8-10 V/cm. If the cooling water is 12-15° the temperature in the gel is kept below 22°. After the desirable separation is achieved the dialysis membrane is removed and the column is placed on a piece of glass tube of the same diameter, as shown in Fig. 2, so that a narrow gap is left between the two tubes. The gel is detached from the glass and now can be moved up and down by means of a supporting piston. The gel is sliced at the appropriate places using a loop of thin wire.

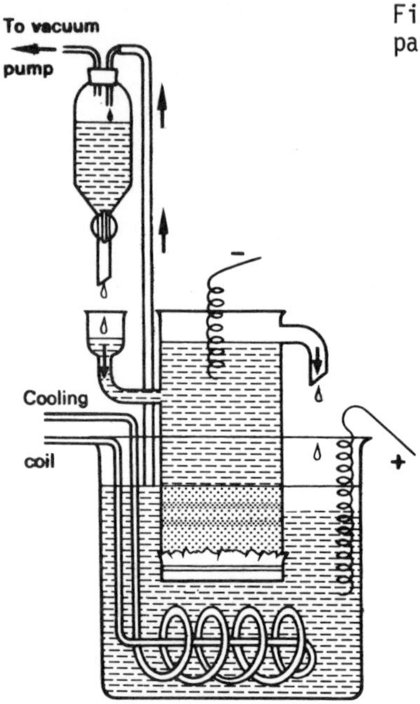

Fig. 1. *(left)* The apparatus for preparative gel column electrophoresis

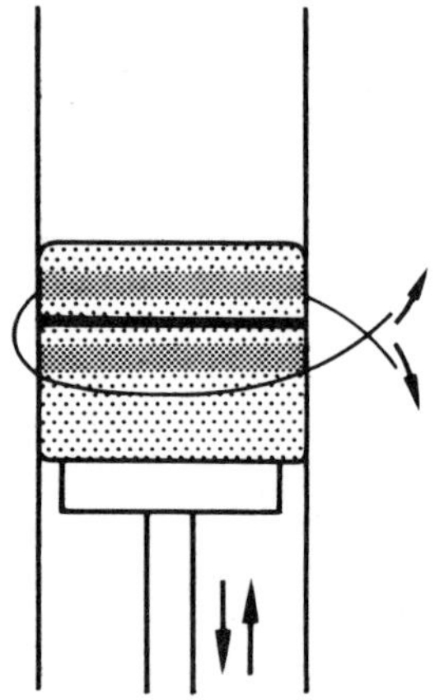

Fig. 2. *(right)* Slicing the gel

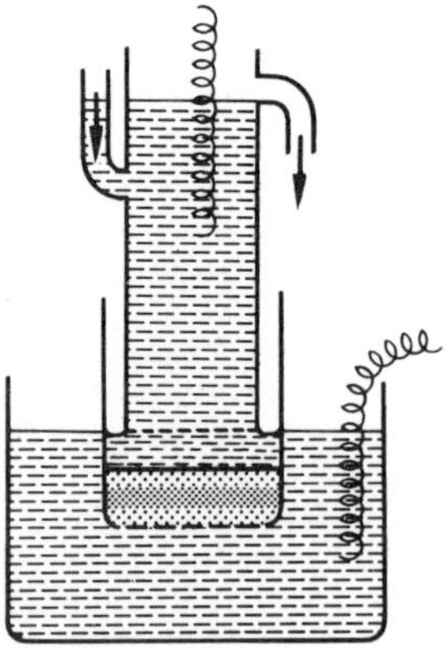

Fig. 3, Set-up for electrophoretic elution of the fractions

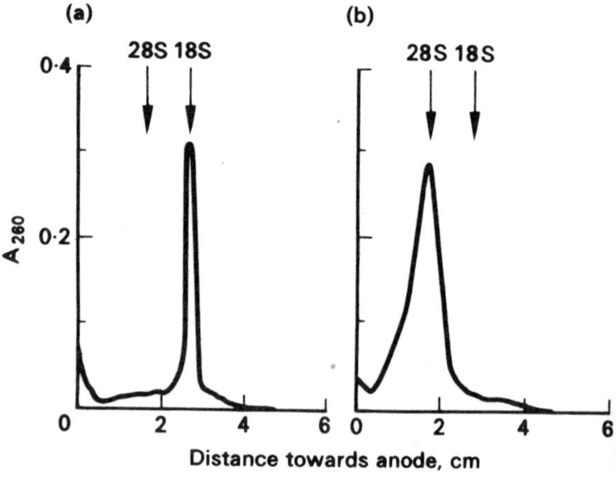

Fig. 4. *Legend opposite*

The slice containing a single fraction could be soaked in buffer for at least 24 h without any loss of material. This helps in removing the low molecular weight impurities. The slice is then placed in a short glass tube whose lower end is closed with a dialysis membrane (Fig. 3). Another narrower tube closed in the same way is inserted in the first one. The space between the slice and the second tube is filled with buffer (2-3 ml) and the current is switched on. The elution takes about 30 min. If the current is left on for 10-15 min longer the material sticks to the membrane of the upper tube and can be taken out and dissolved in a few drops of fresh buffer (additional purification). The yield is usually about 70%. The removal of the dye is achieved by 2-3 precipitations with ethanol (for RNA) and precipitation with 40% ethanol in the presence of 0,008M $MgCl_2$ (5). A re-electrophoresis of 18S and 28S RNA fractions in agar gel (Fig. 4) demonstrates that a clear separation has been obtained.

References

1. Tsanev, R.G., *Biochim. Biophys. Acta* *103* (1965) 374.
2. Dessev, G.D., Venkov, C.D. and Tsanev, R.G., *Eur. J. Biochem.* *7* (1969) 280.
3. Steiner, R.F. and Beers, R.F., *Polynucleotides*, Elsevier, Amsterdam (1961) p.301.
4. Leitin, V.L. and Lerman, M.I., *Biokhimia* *34* (1969) 839.
5. Falvey, A.K. and Staehelin, T., *J. Mol. Biol.* *53* (1970) 1.

Fig. 4. *(opposite)* Electrophoretic profiles of (a) 18 S and (b) 28 S rat liver RNA separated by preparative gel electrophoresis and analyzed by re-electrophoresis in agar gel. *The arrows indicate the positions of the markers run in neighbouring slots.*

6 A TECHNIQUE FOR THE EXAMINATION OF THE RNA OF SMALL AMOUNTS OF PURIFIED RIBONUCLEOPROTEIN PARTICLES

R.H. Hinton
Wolfson Bioanalytical Centre
University of Surrey
Guildford
Surrey, U.K.

and

G.N. Dessev
Institute of Biochemistry
Bulgarian Academy of Sciences
Sofia
Bulgaria

Conditions are described for the concentration of small quantities of ribonucleoprotein particles and examination of their constituent RNA. Protein is dissociated from the RNA by treatment with sodium dodecyl-sulphate (SDS) in low concentration, and the RNA is separated immediately by electrophoresis in an agar gel containing 0.05% SDS. The resolution obtained in equal to that achieved with RNA purified by phenol extraction.

Electrophoresis in agar gels (1,2) is a convenient method for examining the size distribution of molecules in a small sample of RNA. The ease of setting-up the electrophoresis compensates for the slightly poorer resolution as compared with electrophoresis in polyacrylamide gels. However, up to the present time, the technique has only been used to examine samples of purified, phenol-extracted RNA. Such small amounts of native subribosomal particles are normally recovered from sucrose gradients as to make the extraction of the RNA by phenol an undesirable procedure, for the concentration of RNA is so low that a carrier RNA must, in general, be added to assist precipitation, and the presence of the carrier will prevent examination of the whole size spectrum of the RNA molecules in the sample.

Several authors (3-6) have shown that efficient deproteinisation of ribonucleoprotein particles can be achieved merely by shaking with either 0.5% (3-5) or 1% (6) sodium dodecylsulphate (SDS). In particular, Dahlberg and his colleagues (7) deproteinized *Escherichia coli* ribosomes and ribosomal subunits after separation by polyacrylamide gel electrophoresis by soaking the gels in 0.05% SDS and 0.01 M EDTA prior to examination of the RNA by electrophoresis in a second dimension. It therefore appeared worth examining the possibility of dissociating the RNA of mammalian ribonucleoprotein particles directly before separation in agar gels.

MATERIALS AND METHODS

Livers of albino rats were used in these experiments. Polyribosomes were prepared by the method of Leytin and Lerman (8). The livers were homogenized in 0.25 M sucrose containing 0.025 M KCl, 5 mM $MgCl_2$ and 0.05 M Tris pH 7.6. Large particulates were removed by centrifugation for 10 min at 10,000 rev/min in the 8 x 50 ml rotor of an M.S.E.Superspeed 40 centrifuge. The supernatant was decanted and mixed with sufficient 20% aqueous Triton X-100 to give a final concentration of 2%. After allowing the mixture to stand at 0° for 20 min, sufficient 1 M $MgCl_2$ solution was added to give a final concentration of 0.05 M. After a further hour, the polysomes were harvested by centrifugation for 1 h at 17,000 rev/min in the M.S.E. Super Speed 40 centrifuge. The pellet was resuspended using a Teflon-glass homogenizer in 0.05 M $MgCl_2$ containing 0.01 M Tris pH 7.6, and was re-centrifuged for 30 min at 17,000 rev/min. Finally the pellet was re-suspended in about 13 ml of distilled water and dialyzed overnight against 0.01 M phosphate pH 7.6 containing 1 mM $MgCl_2$.

Native ribosomal subunits were prepared from the polysomal suspension by centrifugation for 5 h at 27,000 rev/min of 1.5 ml aliquots layered on 36 ml sucrose gradients (0.5 to 1.0 M) in the buckets of the SW 27 rotor of a Spinco L2 ultracentrifuge. Six gradients were run in parallel, and corresponding regions were pooled. The particles were precipitated by the addition of $MgCl_2$ to a final concentration of 8 mM and ethanol to 41% v/v (9). After leaving overnight at -25°, the particles were collected by centrifugation for 15 min at 4,000 rev/min in a refrigerated centrifuge, and the precipitates dissolved in 0.3-1.0 ml of 0.015 M phosphate pH 7.6 containing 1 mM $MgCl_2$.

Ribosomes containing degraded RNA were prepared by leaving a polysomal suspension, prepared as described above, to autolyze for about 1 week at 0° (10). Under these conditions the RNA chain is broken into a limited number of fragments of a defined size.

In general, electrophoresis was carried out as described by Dessev in the preceding article. The gels (1.25%) were prepared from purified agar (Koch-Light Ltd., Colnbrook, Bucks.) and contained 0.015 M phosphate buffer pH 7.4 and 1 mM EDTA. SDS was normally added to a final concentration of 0.05%. Routinely RNA was released from the particles by the following procedure.– To 0.2 ml of a suspension of ribonucleoprotein particles diluted to give extinction at 260 nm of between 3 and 10 was added 0.02 ml of 0.1 M EDTA pH 7.6. After incubation at 0° for 30 min 0.02 ml of 1% SDS was added. After a further incubation for 30 min at 0° an aliquot of 0.15 ml was loaded onto the electrophoresis plate. Later experiments have suggested that both incubation times may be reduced to about 2 min. Rat liver RNA was used as a marker. Electrophoresis was carried out for 90 min at 220 V, the edges of the plates were sealed with 1.25% agar and the gels were dried in a stream of warm air. When the gel was almost dry, the surface was rinsed with distilled water to remove any incipient crystals of phosphate and SDS. The washing must be carried out with special care in the case of SDS-containing gels, since if these are not dried carefully and well rinsed, they are liable to lift up from the glass plate when dry. The dried gels were allowed to equilibrate with the atmosphere overnight and were then removed from the glass plate and scanned at 260 nm using a recording Spectrodensitometer RSD-220 (Electroimpex, Sofia).

RESULTS AND DISCUSSION

As ribosome subunits migrate electrophoretically in agar gels at rates rather similar to their component RNA molecules (11), some criteria must be formulated to distinguish free RNA from partially deproteinized subunits. Fortunately, the small ribosome subunit migrates rather more slowly than free 18 S RNA. The criteria used for deproteinization of the polysomes were, therefore, that the '18 S' RNA of the experimental specimen should form a sharp peak coincident with the '18 S' RNA in the marker slot and that no protein should be detectable after staining with amido-black (11). Treatment with varying amounts of SDS showed that apparently complete deproteinization could be achieved by incubation with 0.05% SDS as described in the previous Section, provided that 0.05% SDS was present in the gel. If no SDS was added to the gel then treatment with 0.1% SDS was required. In either case, a further increase in the concentration had no effect. Comparison of the resolutions obtained in the presence and absence of SDS in the gel suggested that more reliable resolution of the minor RNA components is obtained if SDS is present. On no account should gels buffered with Tris be used, as this interacts with the SDS and very poor separations are obtained. We found that agar from the source mentioned earlier gave very much better results than another preparation with a similar specification.

The results given in Fig. 1 show the clear resolution of minor RNA species intermediate in size between 18 S and 28 S RNA. These RNA molecules probably derive from 'old' ribosomes which have been partly degraded *in vivo* (12) and not from any degradation *in vitro*. Electrophoresis in polyacrylamide gels shows that the fastest-moving of these minor components can, with difficulty, be resolved into three subfractions (13).

Fig. 2 shows the results of the electrophoresis of RNA liberated from native 40 S and 60 S ribosome subunits, separated as described earlier. The small subunit contains only 18 S RNA as a clearly resolved component. There is also some heterogeneous RNA which may be partly due to non-ribosomal components and partly to degradation products of the 18 S RNA. The large subunit contains 28 S RNA together with some smaller RNA species, presumably the same degradation products as seen in the electrophoretograms of RNA from whole polysomes.

Almost all polysomal preparations are contaminated by traces of ribonuclease which can act on the RNA of the ribosomes themselves. However, incubation of ribosomes for several days at 0°results in the degradation of the RNA reaching a steady state, the RNA having been cut into a number of fragments of distinct size (10). These fragments remain together, held to a very large extent by the secondary structure of the RNA, but may be released by heating the RNA in solution to 85° for 2 min. The RNA of autolyzed polysomes was examined both by the direct technique described earlier, and after extraction with SDS and phenol. The results shown in Fig 3, A and B, reveal a notable difference between the results obtained by the two methods. The 28 S RNA still appears fairly intact in the directly examined specimen but is shown to be almost totally degraded in the phenol-'purified' specimen. However, if the RNA is heated before electrophoresis to reveal 'hidden breaks', then both specimens, as expected, appear to be equally degraded (Fig. 3, C and D), although the presence of ribosomal proteins in the directly examined specimen does seem to have caused some loss in resolution. The explanation for the difference between the non-heated specimens is that the structure of the RNA in the ribosome, especially the 28 S component, is stabilized by the presence of protein. When this is removed, RNA not attached by covalent links will tend to fall apart, a process possibly assisted by the shaking with phenol. In the directly examined specimen, on the other hand, the protein is removed immediately prior to electrophoresis and there is therefore little time for the secondary structure of the RNA to degrade.

From these experiments it may be concluded that electrophoresis in SDS-containing agar gels directly after treatment with SDS has been established as a valid technique for the examination of the RNA of purified ribonucleoprotein particles. Very small amounts of ribonucleoprotein particles can be concentrated by pelleting or by precipitation with Mg^{2+} and ethanol. We found, in the latter case, that satisfactory concentration could be achieved from solutions containing as little as 0.4 A_{260} units of particles per ml. If more than a few percent of extraneous protein is present, a gluey precipitate is formed which cannot be redissolved in the buffer alone but may be redissolved in buffer containing 0.5% SDS. Alternatively, if extraneous proteins are present, the RNA may be released by the addition of SDS to 1.0% and then precipitated with 3 volumes of ethanol (14). When this latter technique is used, satisfactory results are obtained in the presence of a three- or four-fold excess of extraneous protein. Extremely large amounts of extraneous protein do, however, interfere in the electrophoresis, apparently by blocking the entry of the RNA into the gel. Very satisfactory results can be achieved by electrophoresis of a solution containing 5 A_{260} units/ml, and as only 0.15 ml is placed in the electrophoresis slot, this means

Fig. 2 *(opposite)* Electrophoretograms of RNA released from A) native 40 S ribosome subunits, and B) native 60 S subunits, by treatment with 0.01 M EDTA and 0.1% SDS. *The native ribosome subunits were prepared as described in the text. Electrophoresis was carried out as described in the legend to Fig. 1.*

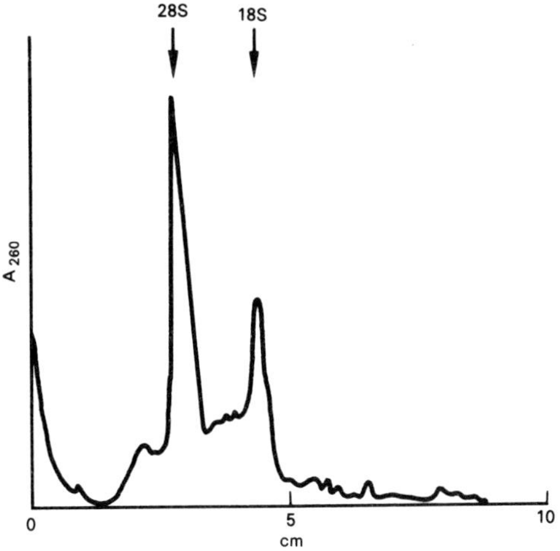

Fig. 1. Electrophoretogram of RNA released from polyribosomes by treatment with 0.01 M EDTA and 0.1% SDS. The treatment was as described in the text. *Electrophoresis was carried out on 0.15 m ml samples for 90 min at 6 V/cm (average measured value) on a 1.25% agar plate containing 0.015 M phosphate pH 7.6, 0.001 M EDTA and 0.05% SDS. Migration was from left to right in the Figure.'A_{260}' values are readings obtained by scanning the dried gel in the spectrophotometer,* not *conventional cuvette readings. The absolute reading for this 'A_{260}' reading on the 18 S peak in Figs. 1 & 2 was of the order of 0.1, the axis extending to A_{260} ∿ 0.4.*

Fig. 2 *(below)*. Legend opposite

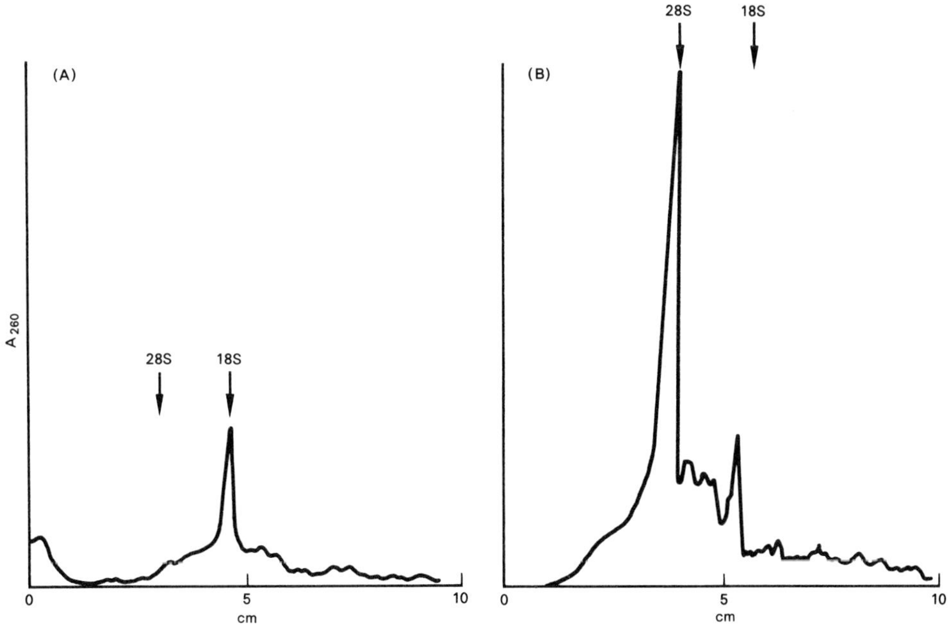

that only 0.75 A_{260} units are required to give a quantitative scan of the spectrum of RNA sizes in a specimen of ribonucleoprotein particles. If the RNA has been labelled, the distribution of label can conveniently be examined by autoradiography (15), acid-soluble nucleotides being removed by washing in dilute acetic acid (11).

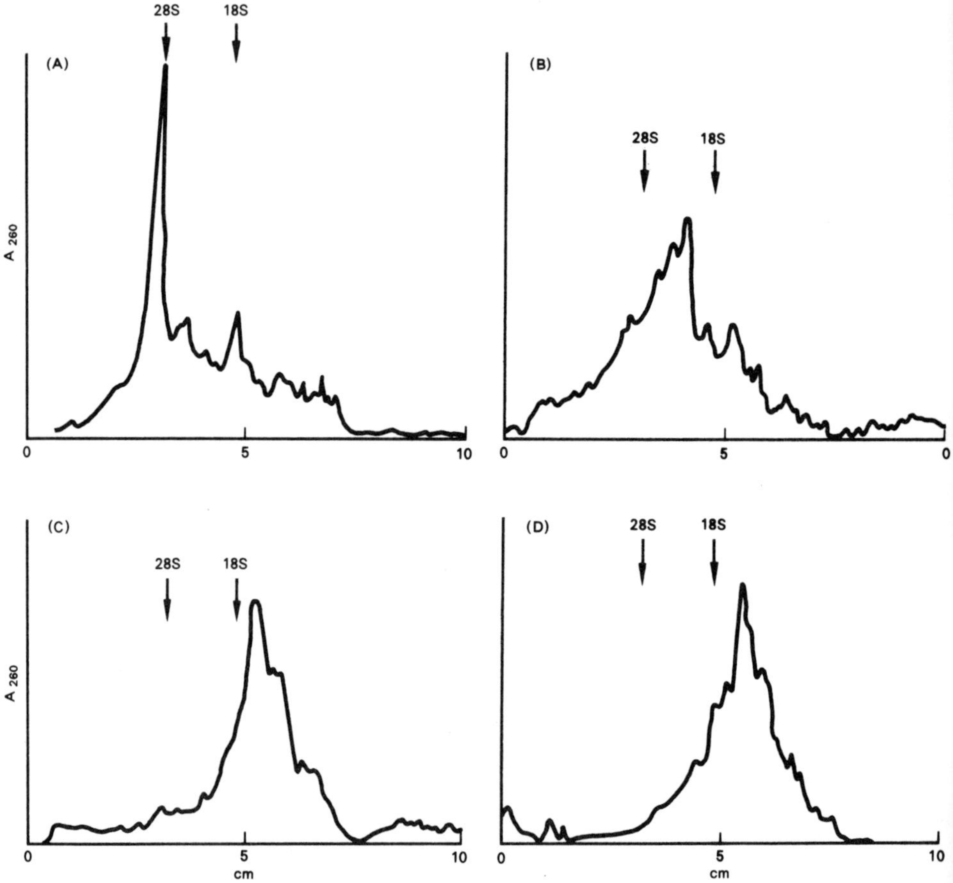

Fig. 3. Electrophoretograms of RNA of ribosomes allowed to autolyze for 1 week at 0°. A) RNA released directly from the ribosomes by 0.01 M EDTA and 0.1% SDS; B) RNA extracted from the ribosomes by phenol; C) as A) but hidden breaks in the RNA were revealed by heating the preparation for 2 min at 85° after release of the RNA by EDTA and SDS; D) as B), but the RNA solution was heated for 2 min at 85° before electrophoresis. *Electrophoresis was carried out for 90 min at 6V/cm (average measured value) on 1.25% agar plates containing 0.015 M phosphate pH 7.6, 0.001 M EDTA and 0.05% SDS.*

Acknowledgements

We thank Professor R.G. Tsanev for his advice and for his hospitality to Dr. Hinton while he was working in the Biochemical Institute of the Bulgarian Academy of Sciences under an Exchange Agreement between the British Council and the Bulgarian Committee for Friendship and Cultural Relations with Foreign Countries.

References

1. Tsanev, R.G., *Biochim. Biophys. Acta 103* (1965) 374.
2. Tsanev, R.G., Staynov, D., Kokileva, L., and Mladenova, I., *Anal. Biochem. 30* (1969) 66.
3. Kurland, C.G., *J. Mol. Biol. 2* (1960) 83.
4. Gilbert, W., *J, Mol. Biol. 6* (1963) 389.
5. Girard, M., Penman, S. and Darnell, J.E., *Proc. Nat. Acad. Sci. (Wash.) 51* (1964) 205.
6. Knight, E. and Darnell, J.E., *J. Mol. Biol. 28* (1967) 491.
7. Dahlberg, A.E. and Peacock, A.C., *J. Mol. Biol. 55* (1971) 1.
8. Leytin, V.L. and Lerman, M.J., *Biokhimiya 34* (1969) 839.
9. Falvey, A.K. and Staehelin, T., *J. Mol. Biol. 53* (1970) 1.
10. Dessev, G.N. and Grantcharov, K., *Life Sci. 9* (1970) 1181.
11. Dessev, G.N., Venkov, C.D. and Tsanev, R.G., *Eur. J. Biochem. 7* (1969) 280.
12. Kokileva, L., Mladenova, I. and Tsanev, R.G., *FEBS Lett. 12* (1971) 233.
13. Ray, C., Agsteribbe, E. and Bont, P., *Biochim. Biophys. Acta 246* (1971) 233.
14. Girard, M. in *Methods in Enzymology* (eds. Grossman, L. and Moldave, K.) *Vol. 12A,* Academic Press, New York (1967) p. 581.
15. Tsanev, R.G., Markov, G.D. and Dessev, G.N., *Biochem. J. 100* (1966) 204.

7 ISOELECTRIC FOCUSING

J.S. Fawcett
London Hospital Medical College Department of Experimental Biochemistry
Queen Mary College
London E1 4NS, U.K.

A brief description is given of the principles of isoelectric focusing (also termed electrofocusing or isoelectric fractionation). Some of the more recent modifications to this technique are then discussed. No attempt is made to review the historical developments or to give a detailed theoretical account of the method. Readers are advised to consult recent excellent reviews by Haglund [1] and by Vesterberg [2], and the papers by Svensson [3-5] who in some really outstanding work developed the theoretical and practical principles upon which this method is based.

Isoelectric focusing is the electrophoretic migration of ampholytes in a pH gradient. Consider a tube filled with a solution of ampholyte and equipped with anode and cathode electrodes (Fig. 1). If a pH gradient is formed with increasing value towards the cathode, the ampholytes in regions more acidic than the isoelectric point will carry a net positive charge and migrate in

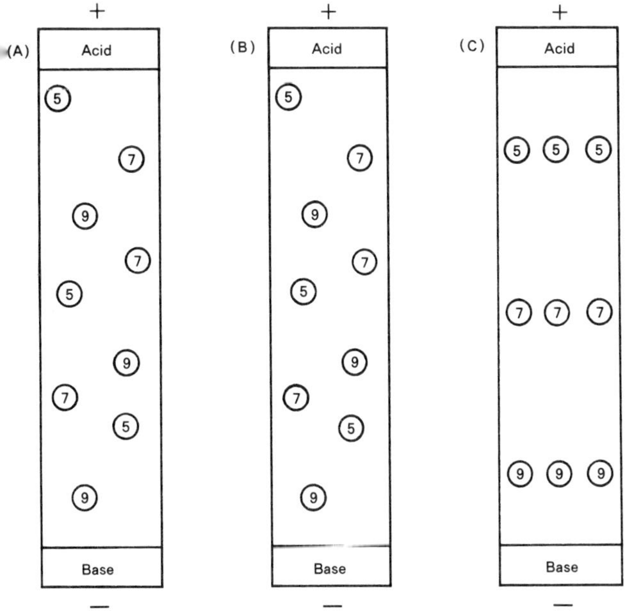

Fig. 1. Principle of isoelectric focusing of a mixture of ampholytes.- Only those isoelectric at pH 5.0, 7.0 and 9.0 are illustrated.

(A) Start of electrofocusing, solution at pH 7.0;

(B) same mixture in early stage of formation of pH gradient;

(C) pH gradient formed and ampholytes focused to isoelectric points.

an electric field towards the cathode. Similarly, ampholytes in a region more basic than their isoelectric point will migrate towards the anode (Fig. 1B). Provided that a stable pH gradient is maintained, an ampholyte will ultimately migrate to the region where the pH is identical with its isoelectric point, where by definition it has zero mobility, and it will concentrate to form a narrow zone. Hence the term 'focusing' (Fig. 1C).

Although the principle of isoelectric focusing has been known for many years, very little progress was made before 1960, due to the difficulty in maintaining stationary pH gradients. Two methods are available for the formation of the pH gradient. One method uses two buffer solutions of different pH which are allowed to diffuse together. Gradients formed in this manner have been called by Svensson "artificial pH gradients". These pH gradients are unstable on passage of an electric current, due to the electrophoretic migration of the buffer ions. They can operate only over short distances.

The second, and more important, method of forming the pH gradient is from a mixture of certain ampholytes, called carrier ampholytes. Here the pH gradient itself is both formed and maintained by the action of the applied electric potential. These gradients have been called by Svensson "natural pH gradients" and are in general very stable. In this system we require a heterogeneous mixture of amphoteric substances having a whole range of isoelectric points. Let us assume that a solution of such a mixture has a pH of 7.0 and is placed in the tube (Fig. 1). To prevent the ampholytes coming into contact with the electrodes where they might undergo oxidation or reduction, the anode is surrounded by a layer of strong acid and the cathode by a layer of strong bases. When an electric potential is applied, all components isoelectric below 7.0 will migrate towards the anode and those isoelectric above 7.0 will migrate towards the cathode (Fig. 1A). In a very short time the concentration of acid components will increase near the anode, while the concentration of basic components will decrease and the pH of this region will fall. In a similar manner the pH will rise at the cathode end. Note that an ampholyte isoelectric at, say, 5.0 and originally near the anode end will first migrate towards the anode, but as the pH of this region falls below 5.0 the ampholyte will change direction and migrate towards the cathode until it reaches the region of pH 5.0.

This procedure of migration and changing pH continues until finally the whole volume between the electrode solutions is occupied with ampholytes arranged in sequence of their isoelectric points (Fig. 1C). The ultimate pH gradient will depend on the distribution of isoelectric points and on the relative concentration and buffering capacity of individual components.

Svensson defined the requirements of these carrier ampholytes to form satisfactory pH gradients. These include good buffering capacity and appreciable conductance at the isoelectric point. For practical reasons they must be readily soluble in water, should be of low molecular weight to facilitate easy separation from proteins, and should be transparent in UV light at 280 nm to permit the detection of protein zones by UV absorption at this wavelength.

A further advance in the technique came with the introduction of synthetic carrier ampholytes. These are polyamino-polycarboxylic acids and were prepared by Vesterburg (6) by the condensation of polyamino aliphatic compounds with certain unsaturated carboxylic acids such as acrylic acid. In this way he was able to make a complex mixture of isomers and homologues of relatively low molecular weight and having isoelectric points distributed over the range 3.5 to 9.5. These carrier ampholytes are now commercially available under the trade name 'Ampholine'*. They are sold in a variety of pH ranges as a 40% solution. Quite recently very narrow ranges extending over only 0.5 pH units have been prepared for limited distribution. It is reported that sometimes aspartic and glutamic acids are added to the Ampholine mixture to extend the pH range at the acidic end, and lysine and arginine to extend the range at the basic end.

In all isoelectric focusing techniques the carrier ampholyte forming the pH gradient and the focused protein zone needs to be stabilized against convection currents. The stabilising systems used so far include density gradient, polyacrylamide gel, granulated polyacrylamide gel or Sephadex gel, and zone convection, and are reviewed in the following pages.

DENSITY GRADIENT COLUMN (Svensson)

The sucrose density gradient method introduced by Svensson has been widely used for preparative isoelectric focusing, and each year an increasing number of highly successful protein separations are reported. The glass column especially designed for this technique is manufactured in two sizes* with operating capacities of 110 and 440 ml. Detailed instructions are given in the manufacturer's manual and will not be repeated here. Further information is obtainable from technical bulletins called *Acta Ampholinae**.

Although the density gradient columns are easy to use, the filling, running and emptying stages must be carefully controlled in order to obtain optimal resolution.

Special density gradient-forming devices* are available for preparing the density gradients. These are designed for loading the column from the top end, and are very convenient to operate. But loading from the top end does not always give a uniform distribution around the annular space, and the sample or the top electrode solution may be unevenly distributed. Skew zones are sometimes obtained. This also applies if the loading method is manual. In our opinion more reproducible results are obtained if the column is loaded from the bottom end. The upward liquid flow ensures a more uni-

* *obtainable from LKB-Produkter AB, S-161 25 Bromma 1, Sweden*

form distribution. The best method for filling the column from the bottom end is to use a 3-channel peristaltic pump (7). This enables the sample to be placed at any position in the density gradient (8,9). One of the channels pumps the dense solution into the mixing vessel which contains light

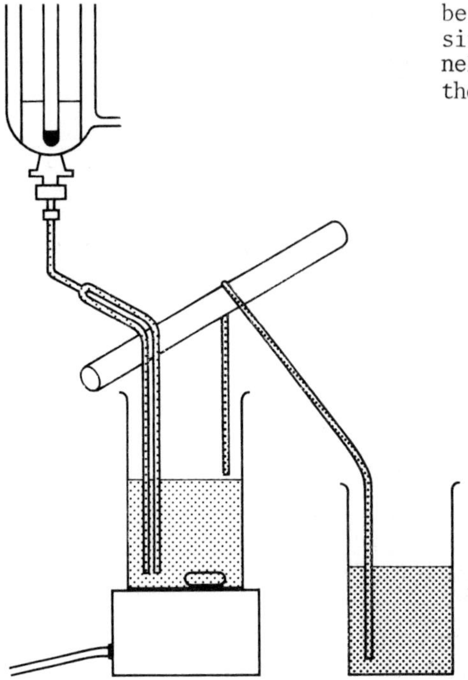

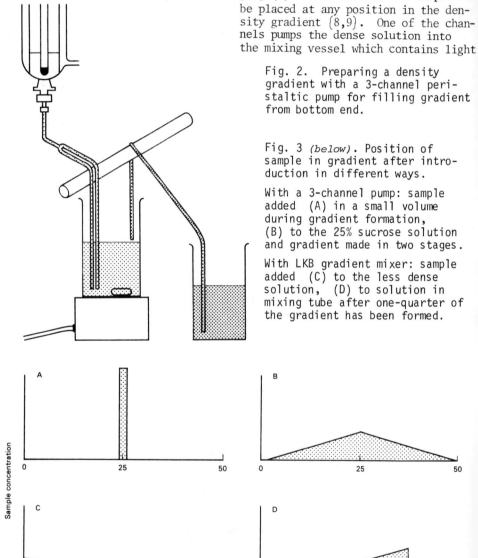

Fig. 2. Preparing a density gradient with a 3-channel peristaltic pump for filling gradient from bottom end.

Fig. 3 *(below)*. Position of sample in gradient after introduction in different ways.

With a 3-channel pump: sample added (A) in a small volume during gradient formation, (B) to the 25% sucrose solution and gradient made in two stages.

With LKB gradient mixer: sample added (C) to the less dense solution, (D) to solution in mixing tube after one-quarter of the gradient has been formed.

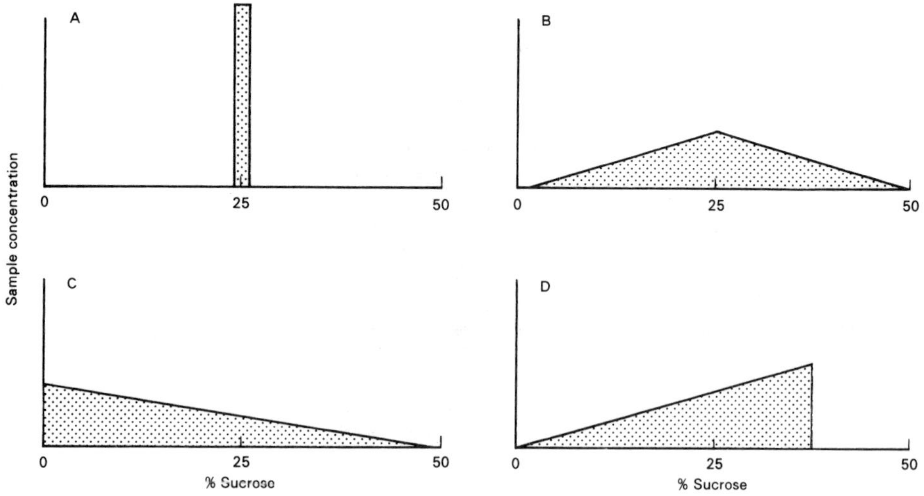

solution, and the other two channels pump from the mixing vessel into the bottom end of the column (Fig. 2). At the desired positions along the gradient the pump is stopped. The sample, in a separate tube, is dissolved in 0.5-1.0 ml of solution withdrawn from the mixing vessel. This solution is then pumped into the column using the two channels temporarily removed from the mixing vessel. Finally, after re-connecting the pump channels, the remainder of the gradient is formed. It is, of course, important that air not be allowed to enter the pump and get blown through the density gradient.

By loading the column in this manner the sample is kept away from the electrode ends (Fig. 3A), where it would be subjected to extreme values of pH and to possible oxidation or reduction at the electrode. Also the duration of the experiment is reduced if the sample is placed near to its final focusing position.

If the sample is in the form of a dilute solution the following procedure is adopted. The sample solution with the addition of sucrose and Ampholine is used to prepare 55 ml of 25% sucrose solution. Also 27.5 ml of Ampholine solution in 50% sucrose and 27.5 ml of Ampholine solution without sucrose are prepared (volumes are given for the 110 ml column). The gradient is now formed in two separate stages: 0-25% sucrose and then 25-50% sucrose. This gives a sample distribution as illustrated in Fig. 3B.

The three-channel peristaltic pump method has many advantages for gradient formation. It may be adapted to suit any particular requirement of sample addition. It is independent of total gradient volume, and gradients down to 10 ml volume have been successfully prepared provided that narrow tubing is used throughout the pump to reduce the 'dead' volume. Linear gradients are formed irrespective of the density of the two mixing solutions.

When the LKB density gradient mixer is used, the method originally given in the manufacturer's manual should not be followed. There it is suggested that the sample be added to the less dense solution. The sample would then be at a maximum concentration at the top electrode end (Fig. 3C) with possibilities of destruction as already discussed. A better method when using the LKB mixer is to add the sample to the mixing vessel after one-quarter of the gradient has been formed (Fig. 3D).

OBTAINING MAXIMUM RESOLUTION WITH THE DENSITY GRADIENT COLUMN

To obtain the maximum resolution for any given protein mixture may require considerable experimentation. The main variables are (1) pH range, (2) concentration of Ampholine, (3) sample load, (4) applied potential, (5) duration of experiment, and (6) isolation of separated fractions. In general a 1% Ampholine solution is used, and a preliminary run in pH 3-10 Ampholine solution will indicate the appropriate pH region for best resolution. To optimize the remaining variables is tedious, and for this reason we developed direct optical scanning methods (see below, OTHER DENSITY GRADIENT COLUMNS) that allow the progress of zone focusing to be observed during the experiment.

Before any further discussion on the best operating conditions we must first consider the position at the focused state. The distribution of protein through the zone is an equilibrium between electrophoretic mobility towards the isoelectric point and simple diffusion away from the point (Fig. 4). The rate of diffusion will depend on the diffusion coefficient D for the protein. It will be reduced by low temperature and high viscosity, but electrophoretic mobility will also be reduced by these conditions.

Electrophoretic mobility will depend on the slope of the mobility against pH plot, $du/d(pH)$, the pH gradient $d(pH)/dx$, and the applied potential E. Svensson has developed the expression for the width of a zone at the focused state as:-

$$x_1 = \pm\sqrt{\frac{D}{E\left(-\frac{du}{d(pH)}\right)\left(\frac{d(pH)}{dx}\right)}} \qquad (1)$$

where x_i is the width of the Gaussian distribution of the focused zone measured from the centre to the inflection point. Sharper zones are therefore obtained by increasing the applied potential, but the gain is only proportional to the square root of the voltage. Higher voltage generates more heat, and the ability to dissipate heat from the column sets the upper limit to any voltage increase. For the 110 ml column the manufacturers have suggested a maximum of 2 watts at the start of the experiment, but it is the condition at the focused state that is more critical. The experiment will start with a fairly uniform conductance through the gradient if a higher concentration of Ampholine is placed in the more dense solution to compensate for the effect produced by the increased viscosity. Once electrofocusing has begun, conductance will no longer be uniform. The components of the Ampholine will become focused, and zones of high resistance are likely to develop when high voltages are used. If the voltage is too great the heat produced in these zones will cause an eddy effect due to the local decrease in density. Layers will mix with those immediately above. It is particularly important to arrange that the zones of low conductance are not in the lower sections of the column where the viscosity from the high sucrose concentration would further reduce the conductance (see manufacturer's manual for polarity of electrodes).

When using the alkaline range (pH 7-10) we have reduced the concentration of Ampholine to 0.5% in the 110 ml LKB column, and have been able to apply final potentials of up to 1,500 V with evidence of enhanced resolution. But by reducing the Ampholine concentration it may also be necessary to reduce the sample load, otherwise the local buffering capacity of the protein might exceed that of the Ampholine.

The use of very shallow pH gradients will increase the distance between zones and will increase the resolution. However, it must be emphasi-

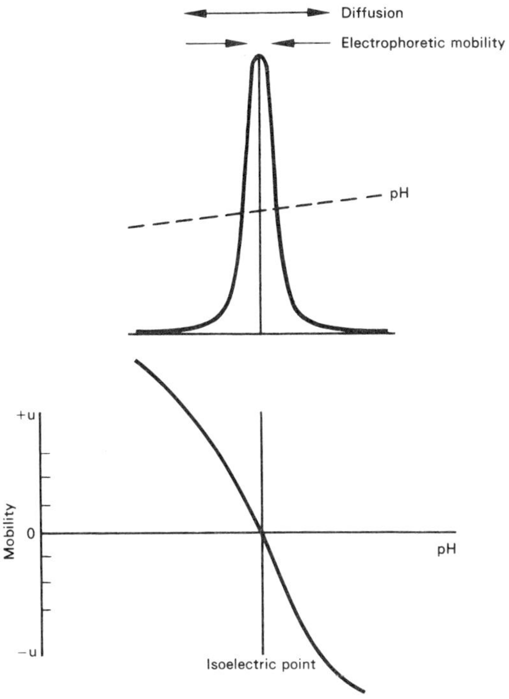

Fig. 4. The focused zone. *Equilibrium between electrophoretic migration towards the isoelectric point and diffusion in opposite direction. Sharper zones obtained by ampholyte with steep plot for mobility/pH.*

zed that a shallow pH gradient will reduce the electrophoretic mobility towards the isoelectric point, and the zones will be wider - see equation (1). Moreover, due to the reduced electrophoretic mobility in the shallow pH gradient it will take longer for equilibrium conditions to be reached.

Excellent resolution was obtained by Albers and Scanu (10) with some polypeptides from serum low density proteins electro-focused in a very shallow gradient pH 4.35-4.85. Two fractions differing by only 0.09 pH units were completely separated free of any detectable cross-contamination.

ISOLATION OF THE FOCUSED ZONES

The best procedure for emptying the column is to pump water at the rate of 1-2 ml per min into the top of the column, and so displace the whole contents downwards and out through the bottom and directly into a fraction collector. If the solution flows through the cell of a UV monitor it may give additional analytical information but this will add to the mixing of the separated zones. Because of the density gradient the liquid must always flow downwards through the flow cell assembly. To minimize mixing, the length of tubing must be kept to a minimum. One troublesome factor is that the dense solution at the bottom end is saturated with gas from the electrode. If the temperature of this solution rises as it passes through

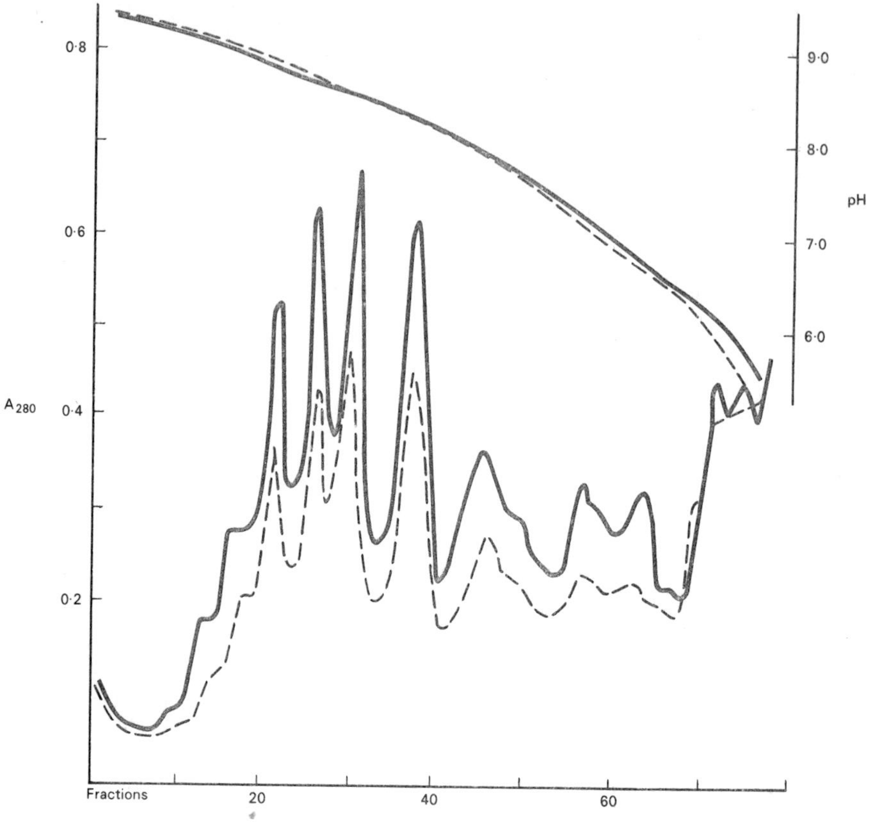

Fig. 5. Isoelectric focusing of pituitary proteins in a 110 ml density gradient column. *Solid line*, 30 mg sample; *broken line*, 23 mg sample. *Conditions: 0.6% Ampholine (pH 7-10); final voltage 1,200 V; run at 4° for 24 h. Note that sharper zones are isolated in the more dense solution that is eluted first.*

the flow cell, some of the gas comes out of solution and collects as a bubble. This bubble remains in the flow cell and can be removed only by interrupting the elution process.

Once the electric current has been switched off, the focused zones will begin to spread by diffusion (Fig. 4) and will continue to spread until they are removed from the column and isolated as fractions. This is especially pronounced with zones that have been focused to very narrow bands by application of high potentials. The extent of the zone spreading can be observed in direct optical scanning experiments (see next Section).

A further spreading of the zone takes place as it moves down the column, and is caused by mixing due to wall effects and turbulent flow in the narrow exit tube. This zone spreading would be reduced by very slow flow rates, but spreading by diffusion would then be greater. Flow rates of 1-2 ml/min are a compromise between the two effects.

Components focused in the bottom section of the column are often isolated as sharper zones (Fig. 5) since (a) they are removed in the shortest time, (b) they have moved less distance in the column and will have less zone spreading due to wall effects, and (c) the higher viscosity of the dense solution will reduce the diffusion rate.

Svendsen (11) has designed an apparatus for isoelectric focusing which allows elution of the column while voltage is applied and thus eliminates zone spreading by diffusion. Unfortunately the fractions are all diluted by a counter-flow of dense sucrose solution containing either acid or base, and the method has little practical value. Continuous flow techniques also allow elution while the voltage is applied, and are discussed in the penultimate Section.

OTHER DENSITY GRADIENT COLUMNS

Recently, Grant and Leaback (12) and Allington and Aron (13) have investigated other types of columns for density gradient isoelectric focusing. These columns as designed by Brakke, Allington and Langille (14) for density gradient electrophoresis are commercially available* in three different sizes. They consist of a vertical cooled column for the density gradient, fitted top and bottom with membranes and electrode assemblies. Facilities are provided for pumping dense solutions into the bottom of the column, thus displacing the density gradient upwards past a UV detector and up into the top section of the column. The density gradient may now be either pumped out through a top adaptor into a fraction collector or, by reversal of pumping direction, returned to the bottom section of the column for further separation.

The smaller model is supplied with a built-in UV monitor. Unfortunately this monitor operates at 254 nm only, and is not ideally suited for the detection of protein zones in isoelectric focusing, as some Ampholine components themselves have considerable absorption at this wavelength. Grant and Leaback have attempted to overcome this difficulty by first measuring the absorption pattern of a pre-focused pH gradient before the addition of the sample and then substracting this from the final result. They achieved good separations of haemoglobin A from haemoglobin F using these columns.

A simplified version of this apparatus (15) designed specifically for isoelectric focusing has been constructed in our laboratory and is

* *from Instrumentation Specialities Co., Lincoln, Nebraska 68504, U.S.A.*

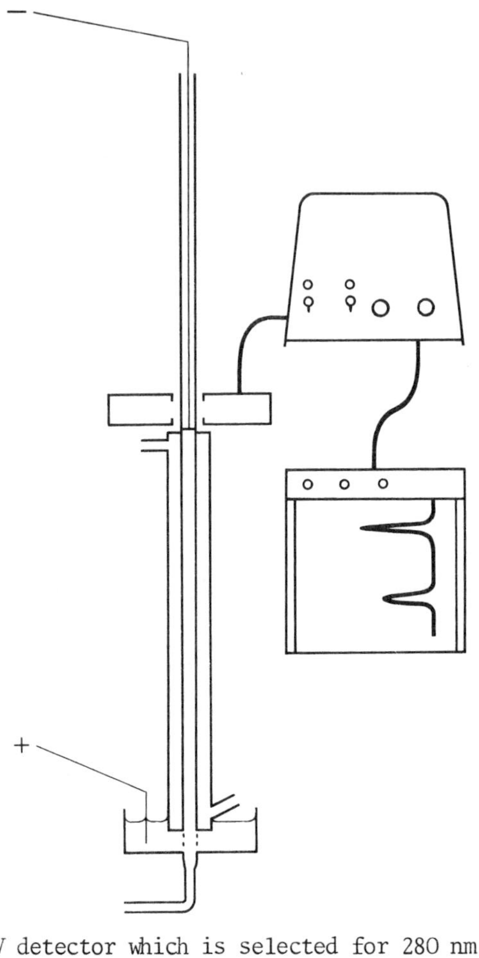

Fig. 6. Isoelectric focusing density gradient scanning column for analytical and preparative use.

illustrated in Fig. 6. It consists of an inexpensive quartz glass tube of 15 mm I.D.* which has the lower half fitted with a cooling jacket. The bottom of the quartz tube is attached to a cylindrical membrane which forms an integral part of the bottom electrode vessel. No membrane is used for the top electrode where a long retractable platinum wire dips directly into the top of the column. The UV monitor is made from a modified detector head from a Uvicord II† by drilling a 2.5 cm diameter hole in the back plate, through which the quartz tube is passed. In this way the UV detector may be moved in succession to scan a number of different columns.

For scanning the column the top electrode is removed, and dense sucrose is pumped into the base of the column to displace the density gradient upwards past the UV detector which is selected for 280 nm absorption. If further electrofocusing is required, dense sucrose is pumped out from the base until the density gradient has returned to its original position in the quartz tube and the top electrode replaced. A typical scan obtained from the isoelectric focusing of horse and whale myoglobins in pH 7-10 gradient is shown in Fig. 7.

Columns of this type have proved most useful in optimising conditions for density gradient isoelectric focusing (see Section on OBTAINING MAXIMUM

* *obtainable from Karel Hackl, 211 Sumatra Road, London N.W.6*

† *manufactured by LKB-Produkter AB, S-161 25 Bromma 1, Sweden*

RESOLUTION), and will find applications in both analytical and preparative techniques.

Other forms of direct optical scanning of isoelectric focusing columns have been devised by Catsimpoolas (16) and by Fawcett (17).

POLYACRYLAMIDE GEL ISOELECTRIC FOCUSING

Isoelectric focusing in polyacrylamide gel was developed in a number of laboratories in 1968, and is now a popular analytical method. Polyacrylamide gel was chosen as the stabilising medium since this material produces negligible electro-osmotic flow. Agarose gel has been used by some workers, but the electro-osmotic flow produced by this material means that stationary pH gradients are not obtained. Even with polyacrylamide gels there is a slow drift of the pH gradient towards the cathode, but this is similar to that observed in the density gradient columns and does not appear to be due to any flow through the gel.

There are essentially two methods of gel isoelectric focusing:- polyacrylamide gel prepared in tubes, similar to disc electrophoresis but longer tubes are generally used, or polyacrylamide gel prepared as a thin film on a glass plate. The latter is a simple but effective method and is most suitable for comparison of a number of samples. Both methods give very good separations and are probably the highest resolving systems available for protein analysis. As an example of the resolution obtained by the thin film technique, Salaman and Williamson (18) were able clearly to resolve proteins with isoelectric points differing by less than 0.02 pH units. Likewise, using the tube method, Hayes and Wellner (19) were able to demonstrate 18 components in their sample of L-amino acid oxidase, whereas only 3 zones could be detected by gel electrophoresis.

Another technique likely to gain popularity is the combination of gel isoelectric focusing in one dimension followed by gel electrophoresis in the second dimension (20, 21). Using this method it was possible to separate about 40 proteins from wheat grains (22).

Practical details of polyacrylamide gel isoelectric focusing will not be given here; they are covered by a number of recent reviews (23, 24). Full details of this thin film method are given elsewhere (18).

SEPHADEX GELS

Isoelectric focusing in Sephadex-stabilised systems was introduced by Radola (25) and by Fawcett (17), but has received little attention. This is surprising as the method offers many attractions for both analytical and preparative purposes. Elaborate apparatus is not required. Sehpadex gels are able to stabilize highly concentrated zones, far higher than density gradient methods (17). It is very simple to isolate the products.

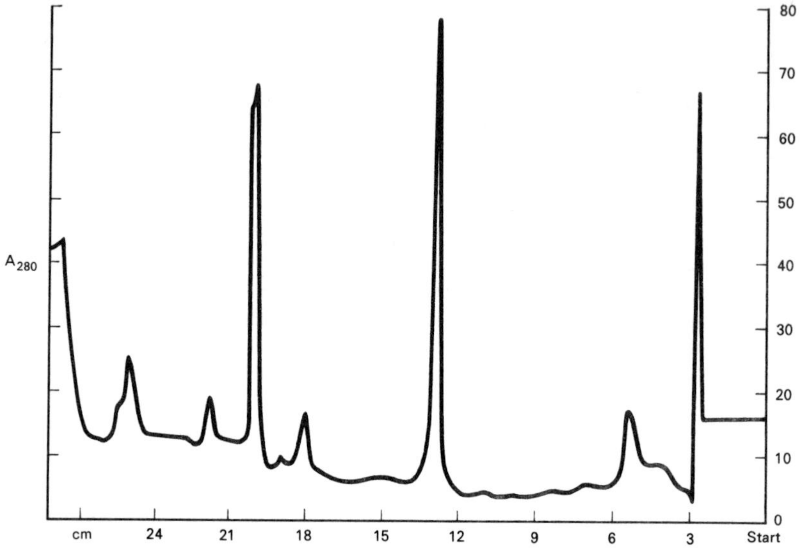

Fig. 7. Absorption scan at 280 nm of a mixture of whale and horse myoglobin (500 mg of each) electrofocused in the gradient scanning column shown in Fig. 6. Cathode at right. *Main peaks from right to left: air-liquid interface, whale myoglobin, horse myoglobin, anode electrode solution.*

Fig. 8 (*opposite*). Zone convection electrofocusing apparatus. Method of Valmet (A) with lid in raised position, (B) assembled for electrofocusing; (C) apparatus based on the design of Kalous & Vacik; (D) polyethylene coil method of Macko & Stegemann.

The apparatus is made from a 10 x 20 glass plate with strips of glass glued round the perimeter to form a trough, which is placed on the horizontal surface of a metal cooling block. A suspension of Sephadex G-75 superfine grade (previously soaked in 1% Ampholine solution of the appropriate pH range) is deaerated by evacuation and spread evenly on the level surface. Filter paper is used to remove excess liquid, and the sample solution is spread on the surface or, when thicker layers of gel are used, is injected into the gel with a fine syringe needle. A filter paper wick, soaked in ortho-phorphoric acid solution and wrapped in cellophane sheeting, connects the 10 cm edge of the gel layer with the anode electrode vessel. Similarly a second filter paper wick soaked in base connects the opposite edge to the cathode. The gel is covered with thin plastic sheeting to prevent evaporation, and run for 18 h at 10-20 V/cm depending on the thickness of the gel layer.

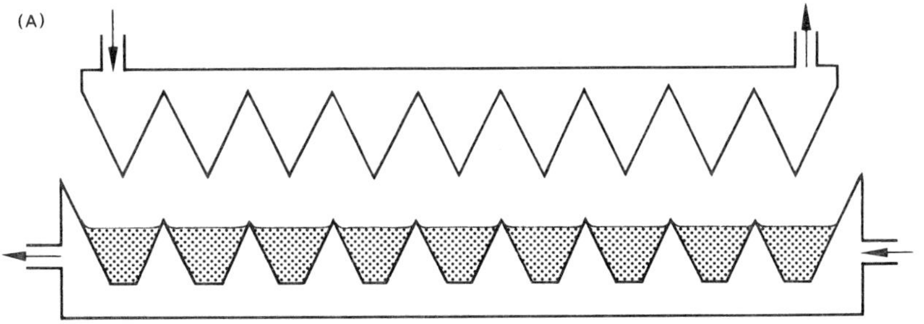

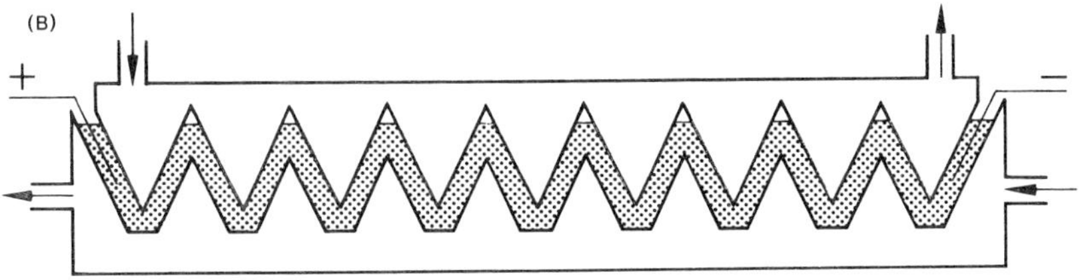

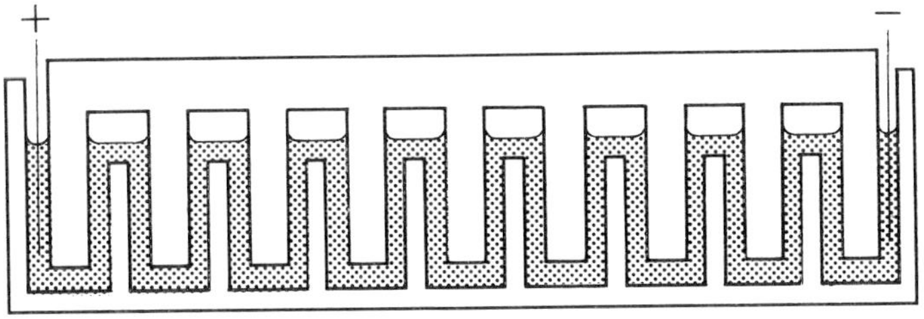

Fig. 8, (A) & (B), *above;* (C), *left;* (D), *below.*

Legends opposite.

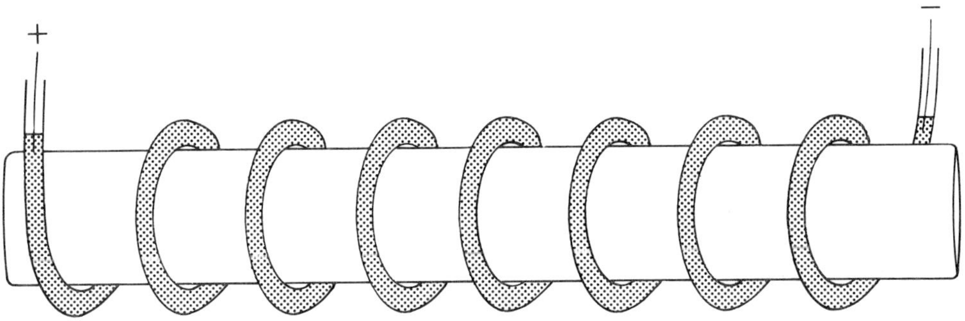

An alternative electrode arrangement preferred by the author consists of two especially prepared electrode vessels. They are rectangular boxes, their sides made of Perspex and their base made from a sheet of porous polyethylene, the pores of which have been plugged with polyacrylamide gel. These electrode vessels are placed on top of the gel bed at each end; the base makes good electrical contact. Electrode vessels of this type have found many useful applications in our laboratory.

After focusing, the zones are detected by the print technique. Filter paper is placed on the surface of the gel, and in a few minutes has soaked up the top liquid layer of gel. The paper is dried, washed with 10% trichloroacetic acid to remove the Ampholine, and finally treated with protein stain. The complete procedure of washing, staining and destaining takes only 30 minutes — very much quicker than the corresponding procedure for thin layers of polyacrylamide gel.

The resolution obtained within the actual Sephadex gel layer is almost as good as that obtained in polyacrylamide gel, but there is some loss in resolution by taking the 'print'. Advantages of the 'print' method are: (a) Ampholine, which would otherwise react with the protein stain, is easily washed away; (b) proteins otherwise difficult to stain can be detected after denaturing on the paper; (c) additional 'prints' may be taken and used for specific staining techniques or for enzyme detection; and (d) the remainder of the material left in the Sephadex layer can be isolated for further analysis.

For preparative purposes the focused zones, located by the 'print' technique, are now cut out of the gel bed and isolated by extraction and filtration. The product will be contaminated with Ampholines and very small quantities of soluble dextrans which have leached out of the Sephadex. Sephadex G-100 and G-200 may be used in a similar manner. They give a more fluid gel layer, and taking the print is more difficult as some of the gel sticks to the paper. This may be overcome by placing two filter papers, one on top of the other, on the gel surface and discarding the paper next to the gel.

OTHER STABILISING SYSTEMS

Hjertén (26) has reported the isoelectric focusing of proteins in free solution using his rotating tube apparatus*. When this equipment becomes commerially available, this technique will provide an extremely useful analytical method, but it cannot be adapted for preparative purposes on a significant scale.

In the near future it is likely that isoelectric focusing will be attempted under zero gravity conditions in space. The absence of gravity forces will provide ideal conditions, and the formation of highly concentrated and very narrow zones could be expected. However, those of us with

* *see Note at end of Art. 4 in this book - Editor*

modest research budgets will have to be content with conditions on earth, and use techniques stabilising against gravitational effects.

Multimembrane compartment cells were used in early isoelectric focusing experiments, and Rilbe (27) has also investigated this method. The main problem is the electro-osmosis produced by the membrane material. This is particularly troublesome in electrofocusing, due to the low ionic strength of the ampholytes at the focused condition. Furthermore, any protein precipitated on the membrane increases the charge on the surface, and the total electro-osmotic flow through the membrane can be so large that the cell is pumped dry.

A more promising approach is the method of zone convection electrofocusing introduced by Valmet (28). His apparatus (Fig. 8A) consists of a shallow rectangular box forming a trough with a ridged base. The underside of the lid has ridges arranged to form a narrow zig-zag channel between the lid and the trough. Both the lid and the trough are cooled by circulating liquid through the hollow cavity. Valmet constructed his apparatus from Perspex with a wall thickness of only 1 mm, and very efficient cooling is obtained. The technique is very simple in operation. First the trough is filled with ampholyte and protein mixture. The lid is lowered and displaces the liquid to give a continuous zig-zag layer (Fig. 8B). Electrodes are inserted at each end, and a high potential applied for several days. A pH gradient is established, and focused zones concentrate at the bottom of the appropriate section. To isolate the zones the lid is carefully raised so that the fractions are trapped in each section.

The advantages claimed for this technique are: (a) substances precipitated at the isoelectric point simply settle in that compartment and do not distribute over a wide zone as sometimes happens in the density gradient column; (b) the method has a high loading capacity, as dense zones can form at the bottom of each compartment; (c) the absence of stabilising media favours the final isolation of the protein; and (d) the rapid isolation of fractions by removal of the lid eliminates any zone spreading by diffusion.

Unfortunately very little information is available on the application of this technique to the separation of protein mixtures, and any assessment of this method is not possible until further reports are published. It is not likely to be a system giving high resolution. In the form described by Valmet the apparatus is very difficult to construct, and at present no commercial models are available.

We have constructed a simplified version of this apparatus which is based on the earlier design of Kalous and Vacik (29). The apparatus illustrated in Fig. 8C is constructed by milling two blocks of Perspex. One block forms the lid, and the other block has strips of 1 mm thick Perspex cemented to its sides to form a long narrow trough 45 cm long, 0.9 cm wide and containing 38 compartments. This gives a working volume of 38 ml.

The apparatus is inserted in a cooling bath and is filled and operated in a manner identical to the Valmet apparatus already described. However, it must be emphasized that only the base and sides of the apparatus are cooled, and the heat removal is not very efficient. It is therefore not feasible to construct a wider version of this apparatus for the separation of larger quantities.

When ovalbumin was electrofocused in this apparatus using a pH 3-10 gradient, it focused satisfactorily and was contained in three adjacent compartments. On the other hand, whale myoglobin electrofocused in a pH 7-10 gradient under similar conditions gave an unsatisfactory result, the components being distributed over several compartments. It is thought that the failure to focus components at the alkaline end of the pH gradient is due to the low cooling efficiency of this apparatus. The pH of solutions of Ampholines in this range is known to be temperature-dependent. If a temperature gradient exists in each compartment then a smooth pH gradient will not be obtained and the focusing process will be disturbed.

An alternative approach to Valmet's zone convection electrofocusing apparatus is the spiral tube as described by Macko and Stegemann (30). Here a plastic tube is wound round a metal rod to form a spiral (Fig. 8D), each turn representing one compartment of the Valmet apparatus. By immersing the unit in a cooling bath and using thin-walled plastic tubing, high cooling efficiency is obtained. After electrofocusing the spiral and its contents are placed in a freezing bath and the fractions isolated by cutting the frozen tube into segments. Quite good separations were obtained by Macko and Stegemann. It is, however, limited to a micro-preparative method, as it is not practicable to use wider plastic tubing. Even with tubing of 0.6 mm outside diameter, not more than 14 coils of the spiral were obtained from 1 m length, giving a path length per coil of 7 cm. Due to the long distance of migration, several days of electrofocusing are required before equilibrium is established. By using special moulding facilities a spiral with more turns could be produced.

Preliminary experiments by the author using coloured proteins in a glass spiral coil with 32 turns (obtained from a broken coil condenser) have given satisfactory results. However, attempts to isolate the zones without cutting the glass spiral have met with failure. A method of removing the contents by slow rotation of the spiral about its axis gave mixing of the zones due to the drag effect of the tube walls.

CONTINUOUS-FLOW ISOELECTRIC FOCUSING

Some progress has been made with the 'scaling up' of isoelectric focusing by using continuous-flow techniques analogous to those already established for conventional electrophoresis. The principle of continuous-flow isoelectric focusing is illustrated in Fig. 9, and is compared with continuous flow electrophoresis. For the electrophoretic method, mixtures to be separated are continuously injected as a narrow band into a flow of electrolyte

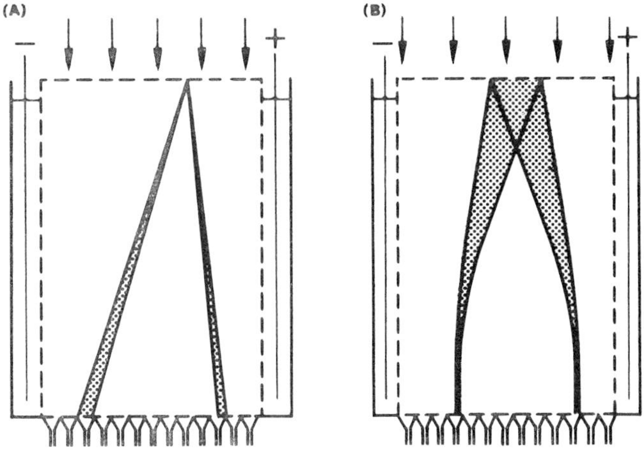

Fig. 9. Diagrammatic representation of continuous-flow - (A) electrophoresis, and (B) isoelectric focusing. Liquid flows at right angles to the electric field.

at right angles to the applied potential. With the isoelectric focusing technique the mixture may be introduced over a wide area. This allows large volumes of relatively dilute solution to be processed. The advantage of the continuous-flow isoelectric focusing technique is that once the components have migrated to their isoelectric point they have zero electrophoretic mobility and will move only in the direction of liquid flow. This means that the position where the zones emerge from the apparatus remains constant during the experiment and is not affected by variations in field strength or liquid flow. This is in complete contrast to continuous-flow electrophoresis where variations of temperature, field strength, conductance or liquid flow all affect the position of the zone at the exit end. In continuous-flow isoelectric focusing the focused zones are isolated while the voltage is still applied. The zones, therefore, unlike those in the density gradient column are not spread by diffusion while the elution takes place (see earlier Section, ISOLATION OF THE FOCUSED ZONES).

Seiler, Thobe and Werner (31), using their version of the Hannig Free Flow apparatus have described some continuous-flow electrofocusing experiments. They have successfully fractionated ox γ-globulin and ox albumin in pH 3-10 gradients. On the other hand, when fractionating haemoglobin or myoglobin in pH 6-8 gradients rather broad zones were obtained. The most likely explanation for broad zones is that the proteins had insufficient time within the electric field to obtain the steady state focused position.

We have investigated two alternative forms of continuous-flow equipment (32) which allow very slow flow rates and give sufficient time for the electrofocusing to take place. One consists of a cooled vertical trough packed with Sephadex gel G-100. A continuous flow of Ampholine solution is pumped to the top of the apparatus, where a horizontal electric field

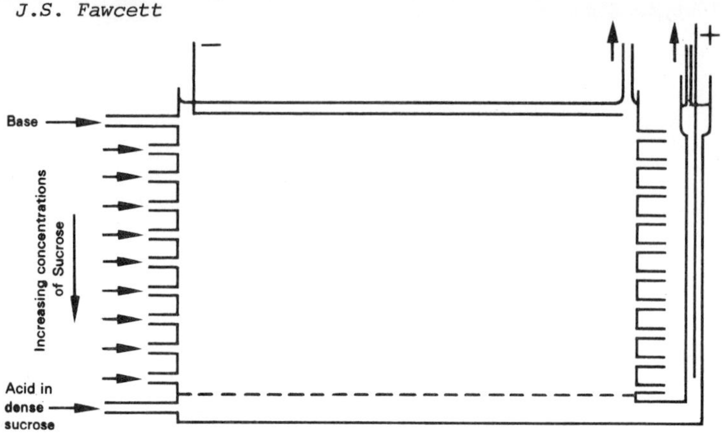

Fig. 10. Continuous-flow density gradient isoelectric focusing apparatus. Cathode along the top, anode in side limb. Density gradient flows from left to right. *(Article 9 shows similar apparatus.)*

focuses the zones as they flow down the bed. Fifty-four collection points at the base of the apparatus are connected to a multi-channel peristaltic pump which regulates the flow through the bed. With haemoglobin solutions and mixtures of other soluble proteins this apparatus has worked well, maintaining focused zones at the same collection point over periods of many days. Difficulties arise with substances that have a low solubility in the region of their isoelectric point. If precipitates coagulate and clog the packed bed, the disrupted liquid flow and electro-osmotic effects produce serious disturbances to the separated zones. In some cases the addition of non-ionic detergents prevents the appearance of precipitates.

The second apparatus uses a continuous-flow density gradient. Philpot (33) as early as 1940 suggested a density gradient for the stabilisation of zones in continuous-flow electrophoresis, and 20 years later this idea was further developed by Mel (34). The isoelectric focusing apparatus (Fig. 10) is constructed from two Perspex cooling plates assembled to form a separation cell 23 cm high, 35 cm long and 0.3 cm wide. The horizontal flow is regulated by a 108-channel peristaltic pump connected to 54 input tubes evenly spaced down one end and 54 exit tubes similarly spaced at the other end. At the input end the top four tubes deliver base solution for the top electrode, and the remaining 50 tubes deliver the sucrose density gradient and Ampholine solution in a series of steps from 1 to 50% sucrose. A small channel along the base of the apparatus is connected with a subsidiary limb containing the anode electrode. Ortho-phosphoric acid in dense sucrose solution is continuously perfused through this channel, which is isolated from the separation cell by a semi-permeable membrane. The acid solution in the bottom channel forms a conducting layer with the anode electrode, and by controlling the concentration of acid the applied potential can be varied along the length of the cell. Because of the high conductance of the Ampholine solution at the input end, a lower potential at this end enables a more uniform generation of heat throughout the cell. Uneven heat generation is liable to disturb the density gradient. This apparatus has given very good results and can separate up to 500 mg of protein mixture per day. To economize on the requirements of Ampholine solution the apparatus was constructed with a relatively small internal volume. Experiments show it could be scaled up considerably without any loss of operational efficiency.

FUTURE DEVELOPMENTS

In the few years since isoelectric focusing has become readily available it has made considerable advances in the fractionation of protein, and this progress is sure to continue. Any prediction as to the way in which this advance will develop is purely speculative.

One would expect the resolving power to increase with the use of very shallow pH gradients and higher potentials, and to place new (and embarrassing) limits for the criteria of purity of proteins. New apparatus would be required with an emphasis on thin film techniques to facilitate the efficient removal of heat. For preparative columns the requirement is for methods of isolating the zones with absolutely minimal mixing.

New carrier ampholytes are likely to appear. In particular, we may expect extensions to the acid and basic pH range of Ampholines, allowing histones and possibly nucleic acids to be fractionated by this method.

Finally, greater use could be made of the isoelectric point value as a characteristic of the protein for identification purposes. This requires greater accuracy and standardisation in the measurement of isoelectric points – a sadly neglected field.

Acknowledgements

The author is indebted to Professor C.J.O.R. Morris who introduced him to isoelectric focusing at an early stage in its development, and wishes to thank him for his continuous advice and encouragement.

References

1. Haglund, H., *Methods of Biochemical Analysis* (D. Glick, ed.) *19* (1971) 1.
2. Vesterberg, O., *Methods in Enzymology* (W.B. Jakoby, ed.) *22* (1971) 389.
3. Svensson, H., *Acta Chem. Scand. 15* (1961) 325.
4. Svensson, H., *Acta Chem. Scand. 16* (1962) 456.
5. Svensson, H., *Arch. Biochem. Biophys., Suppl. 1* (1962) 132.
6. Vesterberg, O., *Acta Chem.Scand. 23* (1969) 2653.
7. Ayad, S.R., Bonsall, R.W. & Hunt, S., *Science Tools 14* (1967) 40.

8. Morris, C.J.O.R., unpublished work

9. Delmotte, P., *Science Tools 17* (1970) 51.

10. Albers. J.J. & Scanu, A.M., *Biochim. Biophys. Acta 236* (1971) 29.

11. Svendsen, P.J., *Protides of the Biological Fluids 17* (1970) 413.

12. Grant, G.M. & Leaback, D.H., *Shandon Instrument Applications 31* (1970).

13. Allington, W.B. & Aron, C.G., *ISCO Appl. Res. Bull., No. 4* (1971).

14. Brakke, M.K., Allington, R.W. & Langille, F.A., *Anal. Biochem. 25* (1968) 30.

15. Fawcett, J.S., *J. Chromatog.,* to be published.

16. Catsimpoolas, N., *Anal. Biochem. 44* (1971) 411.

17. Fawcett, J.S., *Protides of the Biological Fluids 17* (1970) 409.

18. Salaman, M.R. & Williamson, A.R., *Biochem. J. 122* (1971) 93.

19. Hayes, M.B. & Wellner, D., *J. Biol. Chem. 244* (1969) 6636.

20. Dale, G. & Latner, A.L., *Clin. Chim. Acta 24* (1969) 61.

21. Macko, V. & Stegemann, H., *Hoppe-Seyl. Z. Physiol. Chem. 350* (1969) 917.

22. Wrigley, C.W., *Biochem. Genetics 4* (1970) 509.

23. Catsimpoolas, N., *Sep. Sci. 5* (1970) 523.

24. Wellner, D., *Anal. Chem. 43* (1971) 59A.

25. Radola, B.J., *Biochim. Biophys. Acta 194* (1969) 335.

26. Hjertén, S., *Chromatog. Rev. 9* (1967) 122.

27. Rilbe, H., *Protides of the Biological Fluids 17* (1970) 369.

28. Valmet, E., *Science Tools 16* (1969) 8.

29. Kalous, V. & Vacik, J., *Chem. Listy 53* (1959) 35.

30. Macko, V. & Stegmann, H., *Anal. Biochem. 37* (1970) 186.

31. Seiler, N.J., Thobe, J. & Werner, G., *Hoppe-Seyl. Z. Physiol. Chem. 351* (1970) 865.

32. Fawcett, J.S., *Ann. N.Y. Acad. Sci.,* in press.

33. Philpot, J.St.L., *Trans. Faraday Soc. 36* (1940) 38 *[see also Article 8 in this book - Editor].*

34. Mel, H.C., *J. Theor. Biol. 6* (1963) 307.

8 APPARATUS FOR CONTINUOUS-FLOW PREPARATIVE ELECTROPHORESIS

J. St.L. Philpot
formerly *Central Workshop*
Churchill Hospital
*Oxford OX3 7LJ, U.K.**

For the aim of designing an apparatus for thin-layer free-flowing preparative electrophoresis, the initial design was based on the principle of 'transverse' migration stabilized by a density gradient and proceeding normally to the layers [1], as distinct from 'coplanar' migration which occurs parallel to the layers. Separation rate is proportional to migration cross-section divided by migration distance, so that transverse is far faster than coplanar. Ideal resolving power varies inversely as the square root of the 'adiabatic' temperature rise which would occur if there were no cooling. Coplanar migration permits better cooling and therefore greater ideal resolving power for a given actual temperature rise, but its ideal resolving power is somewhat impaired by the parabolic velocity profile. Transverse migration is so fast that there is little time for cooling. It needs only small voltages but large potential gradients which destabilize the flow unless opposed by powerful stabilization, e.g. by a gradient of angular velocity. The advantages of stabilization by angular velocity gradient [2] instead of by density gradient [1,3] are greater effectiveness and avoidance of pollution.

A continuous-flow electrophoretic separator stabilized by a density gradient was made in 1939, but World War II prevented me from perfecting it although I published the theory of it (1). After the war Svensson (4) applied the density gradient to his non-flowing electrophoretic column giving me handsome acknowledgement but remarking correctly that I had tackled too many difficult problems simultaneously. The density-gradient had been applied to the centrifuge by Behrens (5) unknown to me a year before I first used it in electrophoresis. In 1963 Mel (3) had the idea that the density-gradient in Svensson's column could be used in continuous flow. He thereby re-invented the principle of my 1939 apparatus, so the wheel turns full circle.

Soon afterwards I re-entered the field with other ideas. I had never really liked the density-gradient and only chose it for want of a better approach. For a week in 1939 I thought I had avoided it by vertical flow, but the potential gradient made my thin flowing layer of

* *employed (External Staff) by Medical Research Council; now retired at 52 Park Town, Oxford OX2 6SJ.*

migrant look like a row of organ pipes. I interpreted this as an electro-capillary effect, and in 1965 Brown (6) showed that effects like it occur in "uniform" electric fields (including presumably the edges of uniform fields) even in poorly conducting organic liquids where thermal convection is negligible. At potential gradients of 100 or 200 v/cm this electrical destabilisation can be overcome only by powerful stabilization. Vertical flow was used by Dobry and Finn (7), who stabilized it with a viscous polymeric additive, but they did not mention how to separate their additive from proteins. In 1966 I realized (2) that stabilization by a gradient of angular velocity, introduced in 1890 by Couette (8) into viscometry, could be applied to electrophoretic and other separation apparatus. Since then I have had no major problems, only a host of infuriating minor ones, and I now have every hope that the apparatus (Fig. 1) will soon be ready for commercial production.

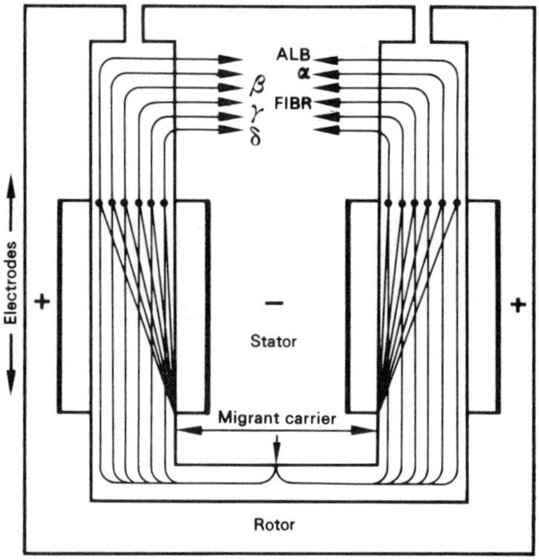

Fig. 1.
Diagrammatic longitudinal section of the rotation gradient apparatus. *A serum protein separation is shown schematically. The 'δ' boundary is an artefact consisting of a stationary raised concentration of carrier electrolyte.*

The most successful continuous-flow apparatus up to date seems to be that of Hannig (9)†, though others based on Mel's (3) have also had some success. My own thinking in 1939 did not go far along the later Hannig route because I decided immediately that, judged solely on speed, it was a dead-end, and I still think so, though it will probably have a permanent advantage in resolving power. The speed is best expressed in the following approximate equation which is a reformulation of one that I gave in 1940 (1)

$$F = \frac{4.186\ ST\mu^2}{\eta\ k}$$

† *For an application of the Hannig apparatus, see Article S-6 in Vol 1 of this series* (Separations with Zonal Rotors) - *Editor.*

where F = ml per min of carrier electrolyte to accommodate migration of fastest component,

$$S = \text{scale factor} = \frac{\text{migration distance in cm}}{\text{area in cm}^2\text{ normal to migration}},$$

T = adiabatic temperature rise,

μ = mobility of fastest component at 0°,

$\bar{\eta}$ = average relative viscosity of carrier electrolyte over actual temperature range, and

k = conductivity of carrier electrolyte at 0°.

In an apparatus where most of the heat is removed by conduction as is customary, the 'adiabatic temperature rise' T has the fictitious value that would result if no heat were removed, which I have calculated is about 2,300° in the Hannig apparatus. This gives a powerful boost to his rate figure ST, but the efficient cooling is obtained at the cost of a long migration path and a small area normal to the migration, which together give his scale factor S the low value 0.05 cm^{-1}. In my substantially adiabatic apparatus T is normally limited to about 20° to avoid protein denaturation, but my S is 2000 cm^{-1}, so that my rate figure ST is about 400 times Hannig's. In both types of apparatus the flow channel is a thin sheet. In mine, which I call 'transverse', the sheet is normal to the potential gradient and the large area is used for supplying current, while in the Hannig apparatus, which I call 'coplanar', the potential gradient is in the plane of the sheet and the large area is used for removal of heat. The ST figures show that for high speed easy supply of current is more important than easy removal of heat. The large area can still be used for removal of heat in a transverse apparatus, as well as for supply of current, and it is so used to some extent in mine, so that ST can be made even larger than the adiabatic figure that I have given. Also T can be doubled to 40° if the carrier electrolyte is precooled to 0° by powerful refrigeration. The theoretical resolving power is given approximately by

$$\mu_r = 5.56 \sqrt{\frac{KD}{T}}$$

where μ_r is the smallest difference of mobility that can be 95% separated, D is diffusion constant and K and T are as before. The scale factor S does not enter into resolving power, but the temperature factor T enters as its reciprocal square root, giving a big advantage to a well-cooled cooled apparatus. Curiously enough, however, published mobility spectra usually seem to show little better resolving power than is theoretically obtainable from an adiabatic apparatus. Some reasons can be offered for this but I will not go into them.

Stabilization by a gradient of angular velocity is illustrated in Fig. 2. The sudden departure of an actual curve from the ideal above a critical voltage is due to electrical destabilization causing turbulent mixing of the deionized water with electrolyte which has diffused from the electrode compartments. The diffusion also causes a slight linear departure by shortening the high-resistance path.

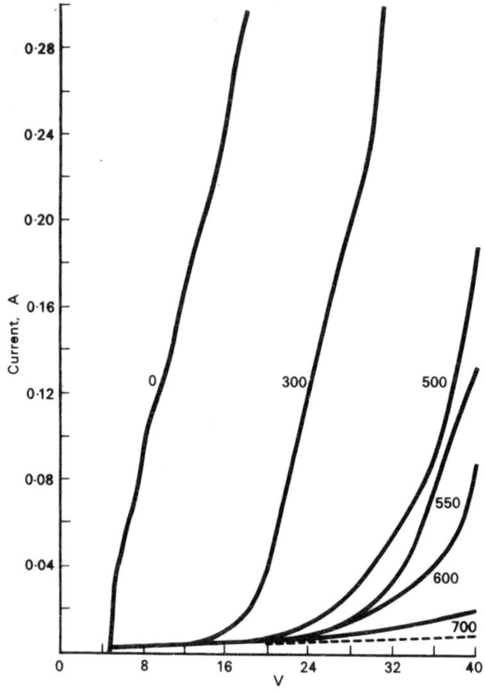

Fig. 2. Stabilisation by a gradient of angular velocity at various values of rev/min. *Current is plotted against voltage (10 V corresponding to 50 V/cm) in a flow-channel of 0.2 cm radial thickness, 23 cm length and 6 cm diameter with deionized water flowing at 1 litre per min. The figures on the curves represent rev/min. The dotted straight line rising to 0.008 amp at 40 V is the ideal curve with perfect stabilization, allowing for the conductivity of the water (10 recip. megohm/cm) and for the back e.m.f. of the electrodes.*

Fig. 3 shows separation of phenol red and naphthol green B by the apparatus, and Fig. 4 shows separation of horse serum proteins, with the albumin stained by eosin. The resolving power is not yet as good as is theoretically possible, through defects which are still being eliminated.

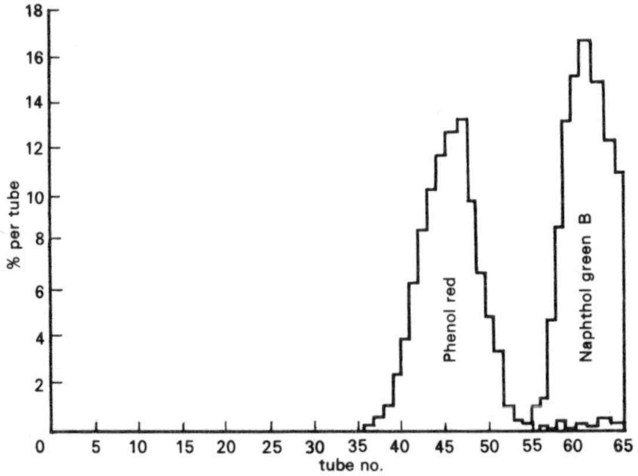

Fig. 3. Electrophoretic separation of dyes. *Migrant flow 2 ml per min of 0.001 M phenol red and naphthol green B. Carrier flow 2000 ml per min of 0.0013 M ammonium acetate + 0.0046 M ammonia; 15 A, 41 V. Concentrations calculated from extinction coefficients at 560 and 690 nm by simultaneous equations.*

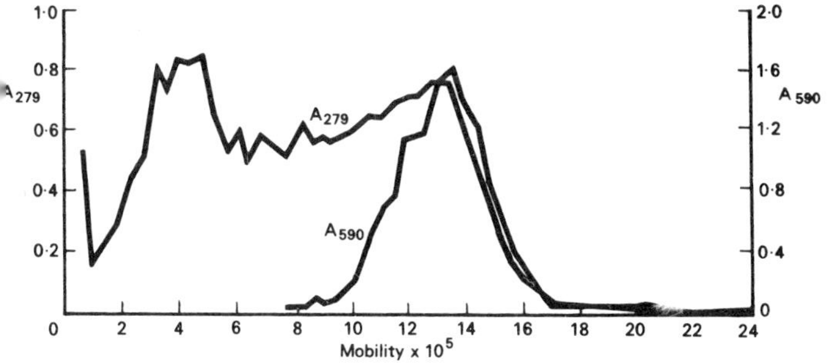

Fig. 4. Electrophoresis of horse serum with eosin bound to the albumin. *Migrant flow 2.2 ml per min of undialysed horse serum plus one-tenth volume of 0.01 M eosin. Carrier flow 200 ml per min of 0.08 M borate. 10 A, 13 V. There was a 4° temperature rise from 20° initially. Readings at 279 nm (protein) and 530 nm (eosin bound to albumin).*

References

1. Philpot, J. St.L., *Trans. Farad. Soc. 36* (1940) 38.

2. Philpot, J. St.L., Brit. Pat. 1,186,184. Appl. date 1966.

3. Mel, H.C., UCRL10640, Office of Technical Service, Dept. of Commerce (1963).

4. Svensson, H., in *Analytical Methods in Protein Chemistry* (P. Alexander and R.J. Bloch, eds.), Pergamon Oxford (1960).

5. Behrens, M., in *Handbuch der biologischen Arbeitsmethoden Urbsn & Schwarzenberg,* (E. Alderhalden, ed.) Berlin (1938).

6. Brown, D.R., *Nat. (Lond.) 202* (1964) 868.

7. Dobry, R. and Finn, R.K., *Chem. Eng. Progr. 54* (1958) 59.

8. Couette, M., *Ann. de Chim. 21* (1890) 28.

9. Hannig, K., *Hoppe-Seyler's Z. Physiol. Chem. 338* (1964) 211.

9 PURIFICATION OF VIRAL SUSPENSIONS BY PARTITION AND BY STABLE-FLOW, FREE BOUNDARY ELECTROPHORESIS

L.C. Robinson
Lister Institute of Preventive Medicine,
Elstree,
Herts, U.K.

In order to concentrate and purify viruses for chemical investigation, large quantities of relatively dilute virus suspensions must be handled. Vaccinia virus, routinely grown in chick fibroblast tissue culture, was first concentrated and partially purified using a partition method. Several different systems of partition in aqueous, two-phase polymer solutions have been investigated [1]. That used in the present work is based on the polyethylene glycol-potassium phosphate system, with scaling up to deal with 20 l batches [2]. After concentration the virus was further purified by one or more cycles of continuous, free-boundary electrophoresis. The electrophoresis cell [3] was modified so that 1.0-1.5 l of material could be purified in 2-3 h the cell being cooled by the continuous pumping of chilled buffers to the electrodes. The virus was separated from contaminants by passage through 12 stable, free-flowing layers of sucrose-buffer.

Stability within the cell is achieved by three factors: (i) *a self-balancing hydrodynamic feed-back;* (ii) *laminar flow;* (iii) *density gradient stability. The virus particle was subjected to a density gradient provided by the layers of sucrose, to a selected pH gradient provided by the buffers, and to the potential applied. The presence of buffers of lower ionic strength on either side of the sample stream produced higher local field strength and enhanced separation. The 'degree of purity' or the gradual loss of contaminating host material during purification was checked by plotting total nitrogen against virus infectivity. In the final stages of purification the loss of host material was measured by complement fixation. Vaccinia virus purified in this way was shown to be free of substances such as copper and FAD [4], once thought to be part of the virus. Also several surface antigens, usually associated with the virus, were removed with no loss of infectivity.*

Viral preparations, grown in cultures of animal tissue, when first harvested have a relatively low concentration of the virus. The bulk of the material consists of host cells and cell debris, together with proteins, peptides and amino acids, derived mainly from the tissue culture medium. The problem of purification consists of the isolation of a very small mass of one component - the virus, from a complex mixture of components, of much greater mass, in a large volume of liquid. The commonly used methods such as differential and density gradient centrifugation are restricted to relatively small samples. Mould (5), reviewing the application of physico-chemical methods to the isolation of viruses, advocated that more use be made of the other physical characteristics of the virus. The concentration and purification of viruses using aqueous two-phase polymer systems was described by Albertsson (1,6). Several viruses have since been purified using a variety of polymer systems. The importance of Albertsson's work was discussed by Tiselius (7) who stressed the simplicity of the partition method.

At the Lister Institute vaccinia virus is now routinely grown in 8-10 1 volumes in monolayers of embryonic chick cells. The virus is first concentrated and partially purified by partition in a two-phase system of polyethylene glycol and potassium phosphate buffer in which the virus partitions at the interface. An earlier version of this method was described by Robinson & Kaplan (2) but the method has since been modified and in its present form operates as a continuous closed system. Although vaccinia virus treated in this way has been shown to be suitable for use as vaccine, further purification was required for biochemical investigations. An electrophoretic method was finally chosen. The first attempt at true free-flow electrophoretic separation was made by Philpot (8) but was not applicable to bulk solutions. Mel (3,9,10) designed an electrophoresis cell based on the principles of stable-flow, free-boundary electrophoresis (STAFLO) which permits the use of bulk solutions. The STAFLO method ensures that free solutions of different densities form continuous moving layers at such a rate that the flow is laminar. Mel's apparatus was used in an investigation of human serum lipoproteins by Tippetts (11) who outlined the potentialities of the method. Some theoretical aspects of the apparatus are presented in the Discussion.

MATERIALS AND METHODS

(a) Partition in polyethylene glycol - potassium phosphate system

Polyethylene glycol (PEG) with an average molecular weight of 6000 was used together with the potassium phosphate buffer described by Albertsson (1). The crude viral suspension was mixed with the phase mixture to produce a final concentration of 20% PEG and 10% potassium phosphate buffer. Volumes of 1 to 2 1 were treated in separating funnels whilst larger volumes (2-8 1) were treated in a continuous and closed system in which each part of the process was an 'in-line' component of the flow system. The interfacial material containing the virus was removed, centrifuged at 5000 rev/min (20 min) and resuspended in a suitable volume of buffer or de-ionized water. The virus obtained in this way was in a highly aggregated form and was homogenized in the cold by ultrasonic treatment.

(b) Purification by electrophoresis

The electrophoresis cell (Fig. 1) was modified for the purification of large volumes of viral suspension (Fig. 1). The cell made from thick sections of perspex cemented together was constructed 2 or 3 times the length of Mel's original design, and the electrode chambers were made larger. Buffer solutions to the electrodes were pumped through by a peristaltic pump (Fig. 1A), the flow rate to each being 10ml/min. The 12 input solutions providing density and pH gradients consisted of sucrose solutions of decreasing concentrations from bottom to top. For most experiments sucrose solutions ranging from 1 to 12% were used, but smaller increments, down to 0.25% sucrose have been used effectively. The input solutions were pumped into the cell by a 12-channel peristaltic pump. The input solutions and the buffer solutions to the electrodes were held in a bath of melting ice in order to protect the virus. The potential applied was usually 100V at about 25 to 50 mA.

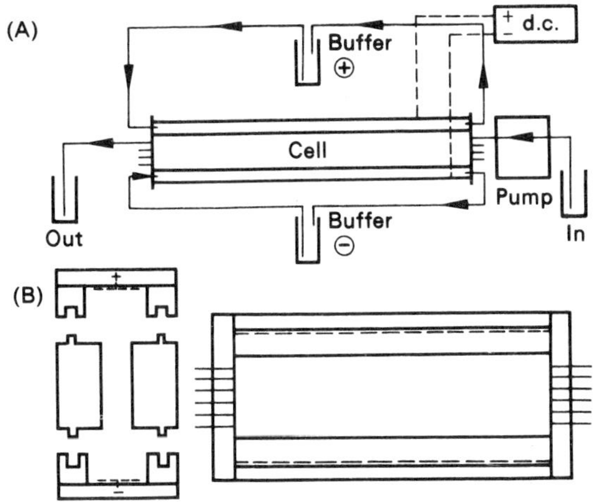

Fig. 1. Apparatus for continuous, free-boundary electrophoresis.
A. Flow diagram; B. Electrophoresis cell.

The input and output ducts were made from sawn-off 16 s.w.g. needles of stainless steel and cemented through the end-pieces of the cell with 'Araldite' strain gauge cement. The platinum foil electrodes, running the whole length of the cell, were cemented in place with 'Bostik' clear adhesive. The electrodes were connected to the terminal by short lengths of 22 s.w.g. platinum wire welded to the foil. The electrode chambers were isolated by strips of Visking dialysis membrane, held in place by the mortice joints machined on the Perspex sides of the cell. The electrode chambers were filled with buffer through the ports and a continuous flow of buffer was maintained (Fig. 1A). Large reservoirs (10-15 l) of buffer were used to minimize the shift of pH. The actual cell (Fig. 1B), limited at the top and bottom by Visking membranes and laterally by the Perspex sides, was 3 cm high and 0.7 cm wide. Each outlet therefore corresponds to a stream of liquid 0.25 x 0.7 cm in cross-sectional area and 60 cm long. The input and output ducts had to be accurately aligned and the cell was levelled on the bench.

The tubing used in the transfer of sucrose solutions to the cell was polyvinyl of 1.0 mm diameter, but the tubing in the peristaltic pump was silicone rubber of the same diameter. Flow rates to the 60 cm cell varied from 1.0 to 2.5 ml/min to each input duct. At higher flow rates some turbulence was sometimes produced in the first 5-10 cm of the cell, but this did not affect the stable flow conditions of the cell as a whole. A slight turbulence in this part of the cell was sometimes an advantage in providing another transport mechanism to aid migration. Short pH gradients were normally used ranging from pH 6.0 at the top of the cell to pH 8.0 at the bottom. In these experiments 0.001 M potassium hydrogen phthalate buffers were used throughout. The virus sample mixed with the appropriate sucrose buffer was pumped into the centre of the cell through input duct No. 7. In addition to the density and pH gradients, streams of different conductivities were used. By having streams of lower ionic strength on either side of the sample stream the higher local field strength enhanced separation, and the more mobile components moved out faster. It was found that efficient spearation could only be achieved if the virus in suspension was not highly aggregated. All virus suspensions were therefore treated with the ultrasonic probe for 3 min at 4°. In all experiments the positive electrode was positioned at the top of the cell.

(c) Virus, and virus assays

The Lister Institute strain of vaccinia virus was used in all experiments. It was obtained from infected monolayer cultures of chick fibroblasts. The virus was assayed by infectivity titrations made by plaque counts on monolayers of chick fibroblasts. Titres were given as plaque-forming units per ml - p.f.u./ml.

(d) Nitrogen estimations

Nitrogen was determined by a direct estimation of ammonia in total micro-Kjeldahl digests by means of the indophenol reaction as modified by Jones (12).

(e) Copper

For many years copper was regarded as an integral part of vaccinia virus, but this has been shown to be not true (4). Copper occurs as a contaminant and is particularly difficult to remove. For this reason the absence of copper was used as a criterion of purity. Copper was estimated by using the apoenzyme of caeruloplasmin with NN-dimethyl-p-phenylene diamine as the substrate. The presence of copper returned the apoenzyme to full oxidase activity. Oxidase activity was measured by a gasometric technique using a modified Cartesian diver (4).

ILLUSTRATIVE SEPARATIONS

(a) Partition of vaccinia virus in PEG/phosphate

Three typical experiments are recorded in Table 1. The results show that by using the partition method a useful level of purification was achieved and at the same time the suspension was concentrated to one-tenth of its original volume. The suspension may however be concentrated still further by reducing the volume of liquid in which the virus is resuspended.

Table 1. Partition of vaccinia virus in the PEG/phosphate two-phase system (p.f.u. = plaque-forming units)

	Before partition	After partition
Exp. TC56		
Volume, ml	1000	100
Titre, p.f.u./ml	9.0×10^6	3.5×10^7
N_2, μg/ml	24.6	4.7
Exp. TC57		
Volume, ml	1000	100
Titre, p.f.u./ml	1.0×10^7	1.0×10^8
N_2, μg/ml	30.0	7.9
Exp. TC58		
Volume, ml	1000	100
Titre, p.f.u./ml	3.0×10^7	2.0×10^8
N_2, μg/ml	25.7	5.6

(b) Purification of vaccinia virus by electrophoresis

Fig. 2 shows the distribution of vaccinia virus after a single cycle through the electrophoresis cell. The suspension of virus consisted of the whole harvest including the chick cells and cell debris. The suspension had a titre of 1.5×10^7 p.f.u./ml, and contained 57 μg/ml N_2. The histogram showed that a partial purification had been obtained, but also showed a very wide distribution of virus. When crude suspension is partially purified by partition the subsequent purification by electrophoresis is much more efficient; this may be seen in Fig. 3.

In this experiment the titre of the viral suspension after partition was 2.0×10^8 p.f.u./ml; it contained 45 μg/ml N_2. The figure shows a considerable reduction in the level of nitrogen in the sample stream with almost no loss of infectivity.

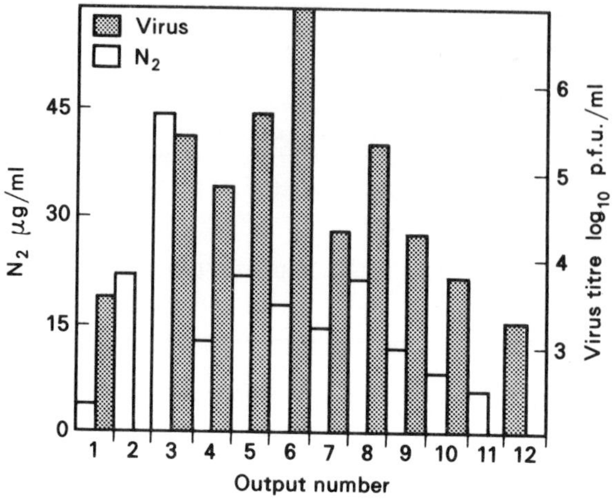

Fig. 2. Electrophoresis of crude viral suspension

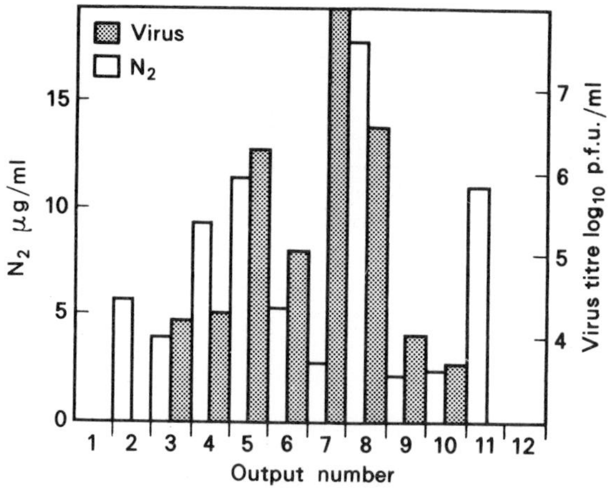

Fig. 3. Electrophoresis after partition

Fig. 4 shows a further gain in purification. In this experiment the virus was recycled through the electrophoresis cell. The result shows that a further reduction in nitrogen occurred with no loss of virus infectivity. Fig. 4 also shows the copper content of each output stream.

DISCUSSION

The partition experiments show that large volumes of vaccinia viral suspensions can be both purified and concentrated. The phase system PEG/phosphate is now used routinely for purifying vaccinia virus for small-pox vaccine. The presence of polyethylene glycol is protective, increasing the stability of viral infectivity. Also the phase boundary of polymer systems has very small interfacial tensions and each phase is rich in water; both factors prevent damage to labile biological materials. Vaccinia viral

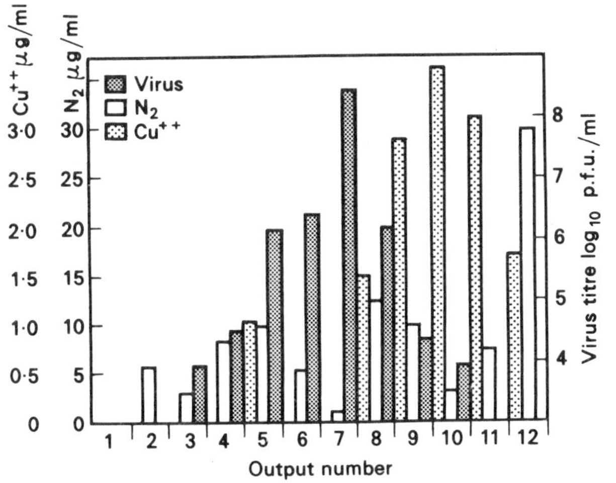

Fig. 4. Electrophoresis after recycling.

particles, concentrated at the interface of this system, must have been highly aggregated, since virtually all the infectivity was sedimented in the centrifuge at 5000 rev/min for 30 min. But after resuspension of the virus, aggregation of the virus particles is reduced quite easily by ultrasonics. Purification with the electrophoresis cell was found to be completely reliable and reproducible, provided that a preliminary purification by partition was carried out first. For high resolution, the fluid flow in the separation cell must be uniform. This is achieved by a self-regulating hydrostatic feedback system provided by the output tubes from the cell. If any fluctuation in flow results in a momentary deviation of flow rate to some of the collection bottles, then the resultant differences in hydrostatic pressure would operate to correct the imbalance.

The positive density gradients, i.e. density increasing in a downward direction in the cell, suppresses convection. The use of appropriate concentrations of sucrose prevents density shifts due to the influx or efflux of molecules into or out of the sample stream, and 'droplet sedimentation' or bombing is prevented (13).

Stability is also dependent on laminar flow within the cell. At least the flow should be laminar for the last two-thirds of the cell. Turbulence at the input end does not upset the overall cell stability. Conditions required for the transition of turbulent to laminar flow are therefore important here. These conditions are expressed in the equation

$Re = \frac{4\rho vR}{n}$, where Re is the dimensionless Reynolds number for closed tubes, ρ is the density, v the linear velocity in cm/sec, n is the absolute viscosity in poise and R is the mean hydraulic radius or cross sect.area/wetted perimeter. If Re is less than 2,100, flow is always laminar; for values above this, the flow may be turbulent, Under the usual operating conditions of the cell Re is about 17 to 18, so that there is a very wide margin of safety for stable flow conditions.

It was found that for similar virus preparations contaminants separated in the electrophoresis cell formed positionally constant peaks. The distribution of nitrogen for example showed four main peaks occurring in outlet duct number 1, number 4-5 and number 11. The distribution of nitrogen in a preparation after two cycles of electrophoresis was however different from that after only a single passage through the cell. The distribution of copper was also constant: a small peak above the virus stream and a large peak below. No single criterion of purity is sufficient to establish the homogeneity of a virus preparation. The estimation of total N_2 is used here merely as a measure of the gradual loss of contaminating host material. Copper is now known not to be an integral part of vaccinia virus, and, since copper is very difficult to remove from proteinaceous materials, its gradual loss was also used as a criterion of purity.

References

1. Albertsson, P.-Å., *Partition of cell particles and macromolecules,* John Wiley, New York, (1960).
2. Robinson, L.C. and Kaplan, C., *Proc. 10th Intl. Congr. Permanent Sect. Microbiol. Standard,* Prague, (1967).
3. Mel, H.C., *J. Theor. Biol., 6* (1963) 159.
4. Robinson, L.C., Thesis, London University (1968)
5. Mould, D.L., *Archs. Biochem. Biophys., Suppl. 1,* 38 (1962) 30.
6. Albertsson, P.-Å., *Biochem. Pharmacol. 5* (1961) 351.
7. Tiselius, A., *Pontif. Acad. Scient. Scr. Var. 22* (1962) 1.
8. Philpot, J. St.L., *Trans. Farad. Soc. 36* (1940) 38.
9. Mel, H.C., *J. Theor. Biol. 6* (1964) 181.
10. Mel, H.C., *J. Theor. Biol. 6* (1964) 307.
11. Tippetts, R.D., Thesis, Ph.D., University of California (1965).
12. Jones, C.R., *Lab. Pract. 16* (1967) 1486.
13. Svensson, H., Hagdahl, L., Lerner, K.D., *Sci. Tools 4* (1957) 1.

10 NEW MATERIALS, ESPECIALLY FOR CHROMATOGRAPHY

A.R. Thomson and B.J. Miles
Chemical Engineering Division,
Building 353,
A.E.R.E. Harwell,
Didcot,
Berkshire, U.K.

The introduction of ion-exchange celluloses and of molecular sieves (Sephadex and polyacrylamide gels) has revolutionized the separation of biologically active macromolecules such as enzymes, and of viruses. Nevertheless, these materials have significant disadvantages which become more apparent on scale-up. Thus the porous nature of celluloses, which is responsible for their high capacity for large molecules, leads to considerable difficulties on the larger scale, as does the compressibility of molecular sieves. In addition, these materials are relatively expensive and have limited stability to chemicals and to microorganisms.

Calcium phosphate and porous silica have also been used, for chromatography and for molecular sieving respectively. However, the physical properties of the available forms of calcium phosphate make scale-up difficult, whilst inactivation of biologically sensitive compounds has been reported on porous glass.

In an attempt to overcome some of the problems encountered in the use of existing materials, porous regularly-shaped particles have been prepared at A.E.R.E. from insoluble inorganic oxides and salts including hydroxylapatite. These have significant advantages for column use since they pack rapidly to form stable, incompressible columns with high flow rates, and can readily be recovered when used for batch separations. Separation based on adsorption and molecular sieving, alone or in combination, can be carried out, and in addition they can be used for concentrating proteins from dilute solutions containing salts. They do not act as substrates for microorganisms, can be regenerated readily with acid or alkali depending on the material, and can be sterilized by heating.

The ideal material for the separation of sensitive macromolecules should meet most or all of the following requirements: it should be stable to pressure and temperature, and to reagents especially those likely to be used for elution, and those required for stability of the species being separated. It should consist of regular particles containing pores large enough to allow the entry of large molecules, it should be available in narrow particle size ranges, and should be as cheap as possible to facilitate its use on a large scale. Finally it should cause minimal damage to the macromolecules with which it comes in contact, and should not be susceptible to attack by microorganisms.

All the existing materials available fulfil some of these criteria, but none are entirely satisfactory.

Fig. 1. *(below)*. Photomicrographs of A.E.R.E. spheroidal calcium phosphate *(left)* and of commercial hydroxylapatite *(right)*.

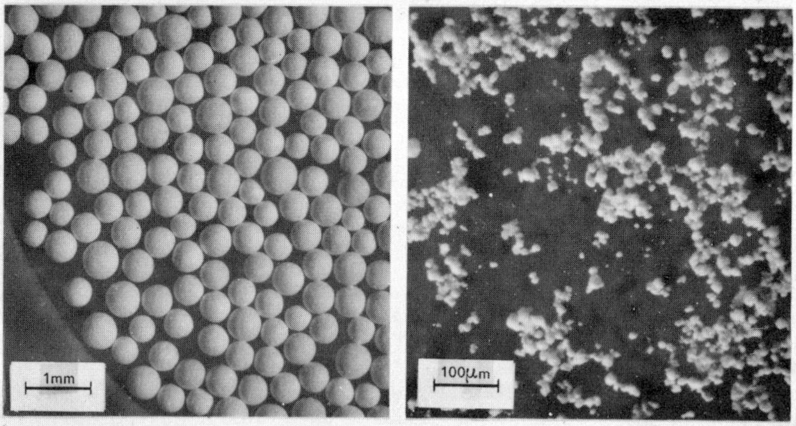

Fig. 3. Photomicrographs of A.E.R.E. spheroidal titania *(left)* and of the starting material *(right)*.

[*For* Fig. 2, *see opposite*

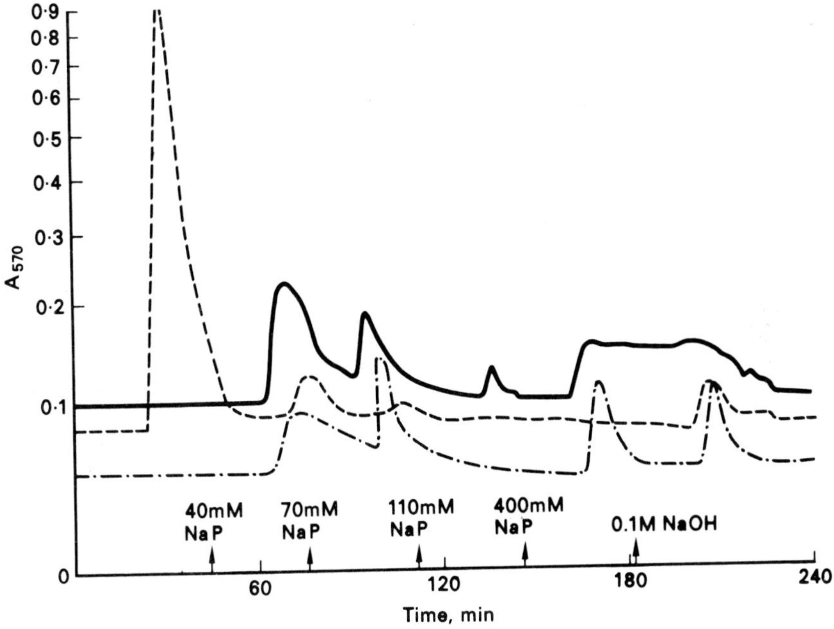

Fig. 2. Separation of bovine serum albumin (BSA) on Bio-Rad hydroxylapatite (———), and on small-pore (-----) or large-pore (-·-·-·) ($Ca_3(PO_4)_2$. Na-P denotes sodium phosphate. *The BSA was loaded in 20 mM sodium phosphate buffer, pH 8.0, at zero time. The last separation was repeated in the presence of 0.9% NaCl, giving an identical trace.*

RESULTS AND DISCUSSION

We have investigated the preparation of inorganic materials in physical forms which would greatly enhance their usefulness especially for separations on a large scale. The best materials that we have developed to meet the above desired requirements are inorganic materials in the form of porous spheroidal particles. A number of these have been made as porous, spheroidal particles in the size range 50-600 µm. These are sized to give particles suitable for use in chromatographic columns. The particles contain pores which are large enough to allow the entry of proteins and polynucleotides, and the pore size can be modified so that molecular exclusion can also be used to enhance resolution. Fig. 1 shows commercially available hydroxylapatite (the crystalline material) compared with spheroidal calcium phosphate prepared by our technique. X-ray scatter diagrams show that the latter is also hydroxylapatite.

In Fig. 2 are shown chromatograms of the same batch of bovine serum albumin on three different hydroxylapatites, one commercially available and two prepared at A.E.R.E. The chromatograms illustrate not only the similarities between our 'large pore' material and Bio-Rad hydroxylapatite, but also the use of molecular exclusion to effect separation (chromatogram on 'small pore' material). As can be seen, there is almost complete exclusion of the albumin from the support in the latter case. It should also be noted that the presence of 0.9% NaCl had no significant effect on the chromatograms. Crystalline hydroxylapatite is fragile, gives relatively low flow rates in columns, and is difficult to re-use. The materials we have prepared, on the other hand, pack rapidly to give stable columns with high flow rates, and can be recycled, *in situ,* many times without repouring. Currently we use an automated chromatographic system in which each sample i loaded automatically, a stepwise programme of buffer changes is carried out and the eluate is analyzed by an automated Folin method. With this system it is clear that our hydroxylapatite can be regenerated by washing with alkali which removes tightly bound protein, that re-equilibration can be effected with low molarity phosphate, and that chromatograms which are reproducible both qualitatively and quantitatively can be obtained over an extended period.

At an early stage we decided to look at titania, expecting to obtain a material relatively inert to proteins. Unexpectedly titania, i.e. TiO_2, is rather adsorptive. Fig. 3 compares particles of porous, spherical titania prepared at A.E.R.E. with the amorphous starting material (laboratory grade TiO_2). These particles settle rapidly to form stable chromatographic beds which give high flow rates (e.g. up to 600 $ml/cm^2/h$).

Chromatography of proteins can be carried out on this material, e.g. Fig. 4 shows separation of a synthetic mixture of bovine serum albumin, γ-globulin, and cytochrome *c*, whilst Fig. 5 illustrates fractionation of an ammonium sulphate fraction from horse muscle. In the latter case there was a four-fold increase in the specific activity of phosphoglycerate kinase (PGK), and 65% of the total enzymic activity was recovered in a single peak In both these experiments, 'large pore' titania was used. In Fig. 6 are shown the results of two experiments with 'small pore' titania. As with our hydroxylapatite, molecular exclusion can be used to enhance the fractionating properties of the adsorbent.

Titania has two further advantages. Firstly, adsorption is not significantly affected by the presence of salts such as NaCl or even 0.3 M $(NH_4)_2SO_4$; thus desalting is often unnecessary before chromatography. Elution can be effected readily, however, with citrate, phosphate, or pyrophosphate depending on the protein. Secondly, dilute solutions of proteins (e.g. 0.1 mg/ml) can be concentrated rapidly, as illustrated for PGK in Fig. 7, and if necessary the adsorbed protein can be fractionated.

Similar results have been obtained with several other oxides. In addition, insoluble salts such as barium sulphate, which is used extensivel for the purification of clotting factors from plasma, has also been prepared in a form suitable for column use.

It should be noted, also that these materials can be sterilized by heating to temperatures of several hundred degrees centigrade.

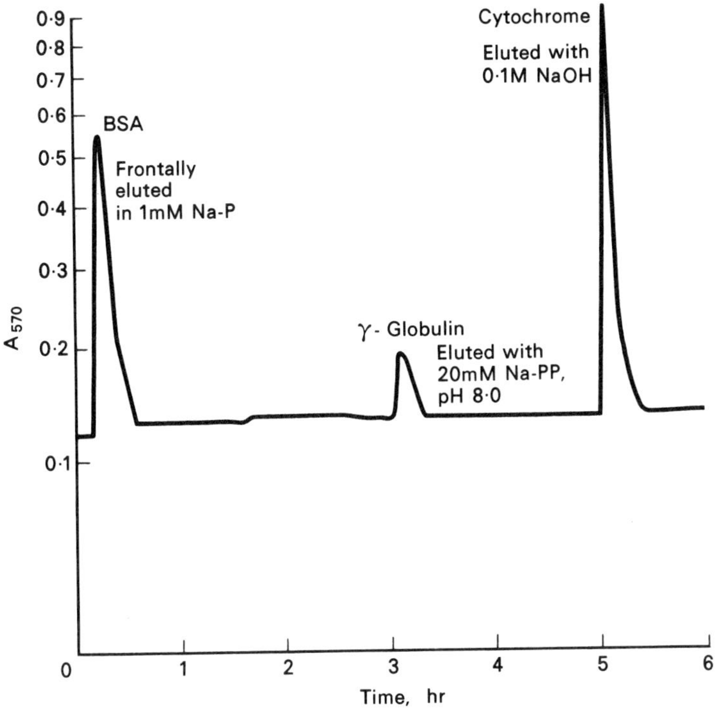

Fig. 4. Separation of a synthetic mixture of proteins on titania. In Figs. 4-7, Na-P denotes sodium phosphate, and Na-PP denotes sodium pyrophosphate. *Column characteristics were initially assayed with individual proteins.*

Acknowledgement

This article appears with the sanction of the Atomic Energy Research Establishment.

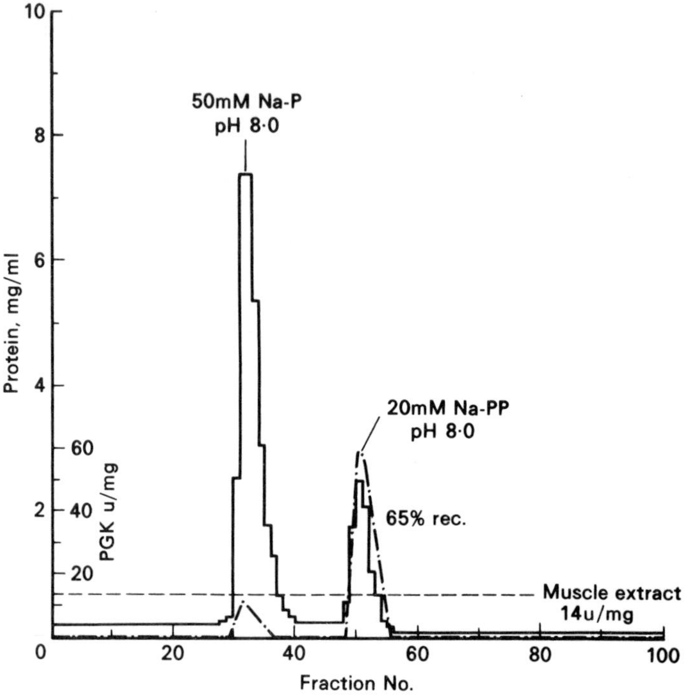

Legends opposite

Fig. 5.

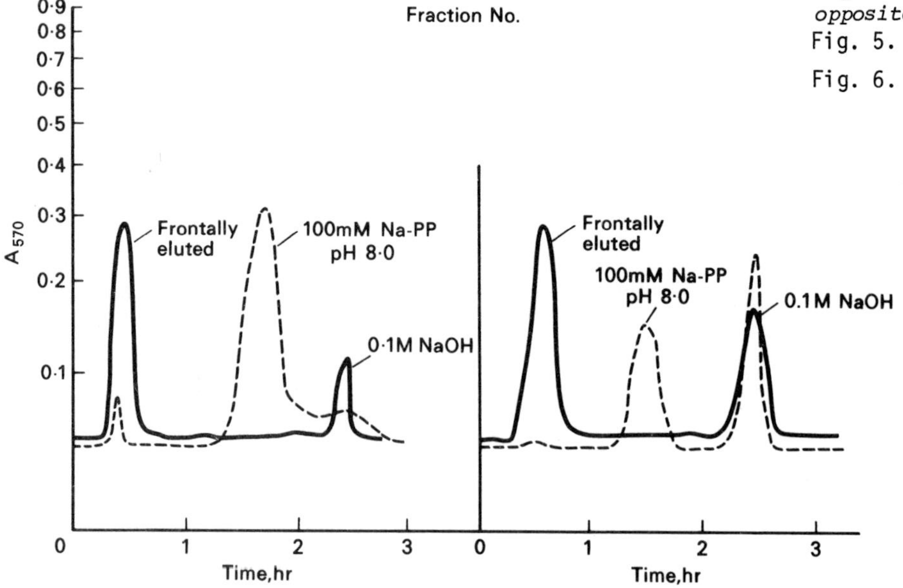

Fig. 6.

Fig. 5 *(opposite, top)* Separation of muscle extract on titania (large-pore size):- ______ , protein, mg/ml; -·-·-·, phosphoglycerate kinase (PGK) u/mg.

Fig. 6 *(opposite, bottom)* Effect of altering porosity of titania:- shown with ovalbumin (left) and myoglobin (right) on small-pore (______) or large-pore (-----) titania. *Each protein was equilibrated and loaded in 20 mM Tris, pH 8.0, at zero time.*

Fig. 7. Concentration of phosphoglycerate kinase on titania. *A dilute solution (0.1 mg/ml in 10 mM ammonium acetate, pH 8.0) was loaded at zero time on a 1 x 50 cm column of large-pore TiO_2 at 600ml/cm^2/h.*

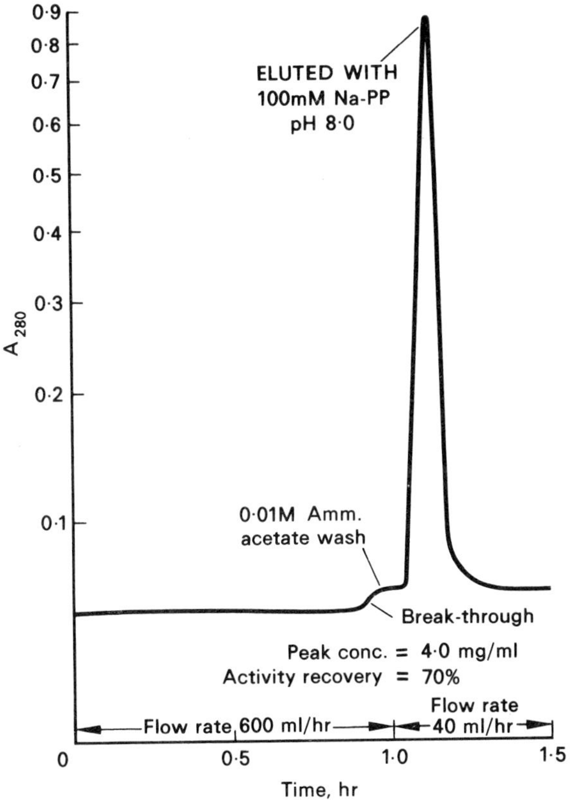

11 THE USE OF AFFINITY CHROMATOGRAPHY IN NUCLEIC ACID BIOCHEMISTRY PARTICULARLY FOR HEPATIC DNA POLYMERASE PURIFICATION

I.R. Johnston, M.E. Haines and A.M. Holmes
Biochemistry Department
University College
London WC1E 6BT, U.K.

Literature on applications of affinity chromatography in the nucleic acid field is surveyed. An account follows of the authors' work on the soluble and nuclear DNA polymerases of rat liver. The nuclear enzyme binds relatively tightly to DNA-cellulose columns. Use has been made of this fact in its purification. The enzyme appears (on G-100 Sephadex) to have a molecular weight of 60,000 and when about 400-fold purified shows a single band of about 30,000 m.w. by SDS-gel electrophoresis. Under a variety of conditions the soluble DNA polymerase seems to bind less tightly to DNA-cellulose.

Nucleic acids have been immobilized in or on insoluble supports by a variety of methods such as molecular sieve entrapment in polyacrylamide, agarose, agarose + polyacrylamide or Sephadex gels; by drying on to cellulose or nitrocellulose, or by UV-irradiation following drying on to cellulose or polyvinyl alcohol [refs. in 1-3]. Attachment of nucleotides, oligonucleotides and polynucleotides specifically through the formation of a phosphodiester link has been achieved using carbodimide reagents both in aqueous and non-aqueous conditions. Using dicyclohexylcarbodiimide, T_4-DNA (via its glucosylhydroxyls) and polyribonucleotides have been similarly attached to acetylated phosphocellulose [see 4 for refs]. More recently cyanogen bromide has been used to couple nucleic acids [5] and nucleotide derivatives [6,7] to agarose; tRNA, periodate-oxidized at its 3'-OH end, has also been linked to aminoethylcellulose [8]. Apart from studies on DNA polymerase and ligase, use has been made of some of these materials in affinity chromatography in at least three ways. First, in the fractionation of oligonucleotide and nucleic acid mixtures [e.g. 9, 10]; second, in studies on the interaction of antibiotics with DNA [11]; and, third, the most extensive application, has been in the preparation of proteins involved in nucleic acid synthesis and degradation. These include a factor regulating transcription of ribosomal genes [12], T_4-gene 32 protein [1,2], Staph. aureus *nuclease [6], DNA polymerases I (2) and II of* E. coli, *and DNA polymerase of* Micrococcus luteus *[13] and RNA polymerase of* E. coli *[see 1 and 2 for refs.].*

To achieve high purification of bacterial DNA polymerases in partially purified form, use has been made of columns of DNA bound to cellulose

(Micrococcus [13]), embedded in a gel matrix such as agar (*E.coli* DNA polymerase II (14) or polyacrylamide plus agar (*E.coli* polymerase I(2)). Apart from calf thymus terminal transferase (15), mammalian DNA polymerases have not been as extensively purified. Rat liver nuclear DNA polymerase is of particular interest since it occurs predominantly in nuclei not involved in DNA synthesis (16). The present work deals with an attempt at its purification using DNA-cellulose columns.

METHODS

DNA-cellulose was prepared as described by Litman (13). Calf thymus DNA, Sigma Type 1 (2 mg/ml in 0.01 M NaCl, 0.01 M Tris-HCl, pH 7.5) was mixed with, and then dried on to, Whatman CF2 cellulose powder (previously washed in N HCl, 5 mM EDTA, distilled water and then dried at 37°). Of this solution, 5 ml was used per g cellulose powder. Usually about 5-6 g of cellulose was processed at a time, the slurry being evenly spread round the walls of a 500 ml beaker and dried using a cool air blower. Drying was completed by storing overnight at 37°. The resulting material was stirred for 20 min in portions of 1.5 g in 25 ml absolute ethanol whilst exposed to UV light at 254 nm from a Phillips 6W-TUV lamp placed 11 cm away. The material was sucked dry and washed free of unbound nucleic acid with 10 mM NaCl. The DNA-cellulose, after suspension in 0.05 M Tris-HCl, 1 mM EDTA, pH 7.5, was used as such in an 8 x 1.3 cm column or was activated with DNase I (17) before use.

Nuclei were prepared from rat liver using a 2.4 M sucrose step (16) and non-histone protein extracted (18). All buffers contained 20% w/v glycerol, 1 mM EDTA, 1 mM β-mercaptoethanol or dithiothreitol.

DNA polymerase was assayed in 0.125 or 0.25 ml incubation mixtures. The following were contained in 0.25 ml; Tris-HCl, pH 7.5, 15 μmole; $MgCl_2$, 2.5 μmole; dATP, dGTP, dCTP and ^{3}H-TTP (40 μCi/μmole), 25 n-mole each; 2-mercaptoethanol, 0.25 μmole; activated calf thymus DNA, 50 μg; bovine serum albumin, 25 μg; and enzyme protein. After incubation for 1 h at 37° the reaction was stopped and the radioactive material processed and counted for 10 min as before (16), except that precipitates were collected by filtration on Whatman GF/C circles.

RESULTS

During the preparation of the non-histone protein fraction, dialysis of the M-NaCl extract against 0.15 M NaCl causes precipitation of nucleohistone, thereby removing the greater part of the nucleic acid. However, unless the bulk of the remainder was also removed, 25-50% of the DNA polymerase activity came straight through the DNA-cellulose column. Neither treatment of the extract with DNase, RNase or both together, nor the use of the phase separation technique (19) was effective - the latter procedure being attended by large losses of activity. Finally the use of Sepharose 6B columns run in 2 M NaCl gave complete recovery of enzyme and removed most of the nucleic acid (Fig. 1).

The resulting enzyme binds completely to DNA-cellulose at an ionic strength of 0.2. Either a linear gradient of NaCl (Fig. 2) or stepwise elution can be used for removal of the enzyme. Nuclear DNA polymerase elutes from the column in the ionic strength range 0.65-0.95. (Using a phosphocellulose column at the same or at a lower pH of 6.4, elution of the enzyme is complete at about an ionic strength of 0.5.) In a typical experiment on DNA-cellulose, 45 mg of protein from the Sepharose 6B peak (about 10-fold purified) was loaded and yielded about 350 µg of protein in the DNA polymerase peak. Recovery of enzyme is 60-90% of that loaded, so that a substantial purification is achieved at this step. This is also revealed by SDS-polyacrylamide gel electrophoresis which shows only about seven polypeptide bands. Further purification can be achieved on QAE-Sephadex at pH 9.0, I = 0.1, the flow-through peak almost completely eliminating all bands except one of mol. wt. 30,000 (SDS gels). Since the enzyme has a mol. wt. of 60,000 on G-100 Sephadex it would appear to consist of two sub-units.

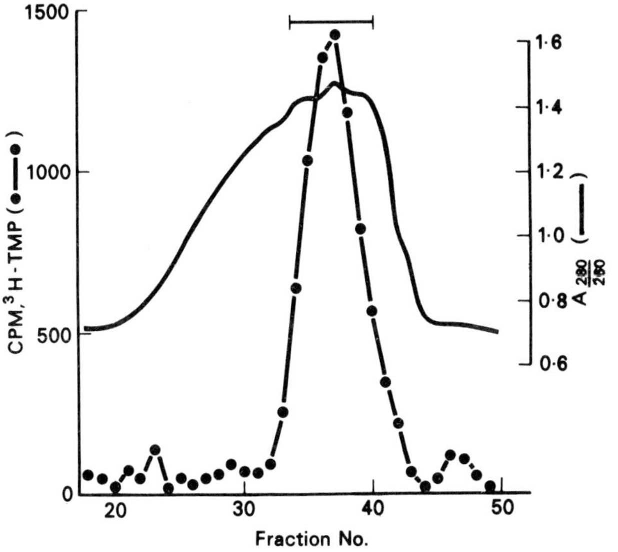

Fig. 1. Removal of nucleic acid from non-histone protein fraction. *A Sepharose 6B column 1.8 x 32 cm equilibrated with 2 M NaCl in 0.05 M Tris-HCl, pH 7.5 was used, and 80 mg non-histone protein loaded. Fractions, 4 ml. Assays were conducted as previously described* (17).

Attempts to purify the soluble DNA polymerase of rat liver using rat liver or calf thymus DNA-cellulose columns have yielded a degree of purification smaller than that obtainable on a phosphocellulose column (50-fold). This stems from the weaker affinity of the soluble enzyme for DNA-cellulose compared with the nuclear enzyme (Table 1). Lowering of the pH to 6.5 and ionic strength to 0.05 brings about binding of the bulk (90%)

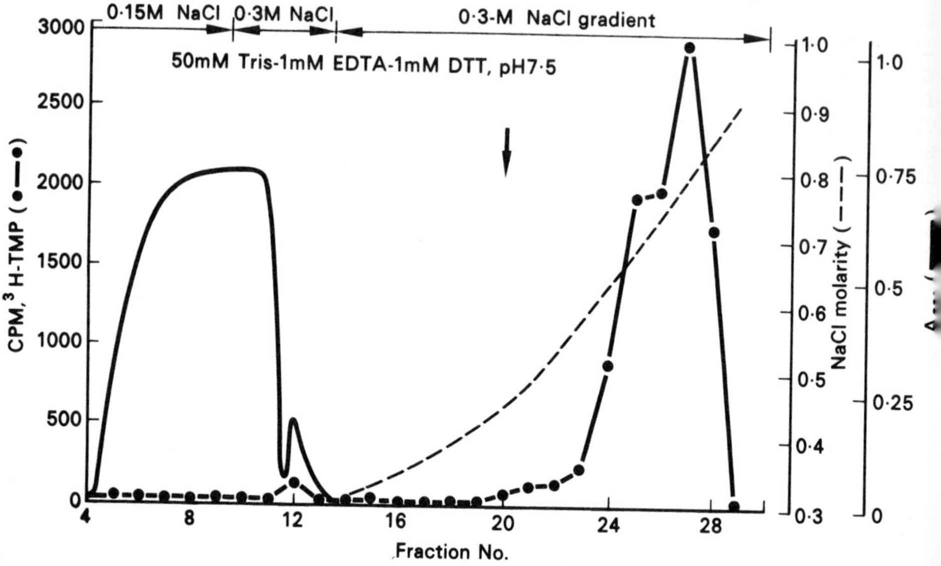

Fig. 2. Purification of rat liver nuclear DNA polymerase on a column of calf thymus DNA-cellulose. *With a 2 x 8 cm column, 45 mg protein (136 units) from a Sepharose 6B column was eluded and eluted as shown. One unit = incorporation of 1 n-mole of* ^{3}*H-TMP per h.*

of the soluble polymerase activity to the DNA-cellulose column. Attempts to elute the polymerase activity after the removal of the bulk of the protein from the column lead only to relatively small increases in purification (e.g. Fig. 3). The DNA cellulose here appears to behave more as an ion-exchange column. The omission of EDTA from the buffers or the addition of Mg^{2+}, or the use of starting fractions of greater purity, or the use of non-irradiated DNA bound to G-200 using a water-soluble carbodiimide (8), has not enabled us to substantially improve the purification of soluble polymerase activity. Further, it now seems possible that all three peaks seen in Fig. 3 may be distinct enzymes.

CONCLUSIONS

The use of DNA immobilized on solid supports has enabled considerable purification of several DNA polymerases and associated proteins. Recent work on the binding of natural and synthetic RNA molecules to Sepharose 4B (5,8,20) can now be expected to facilitate the purification and study of specific RNA-binding proteins.

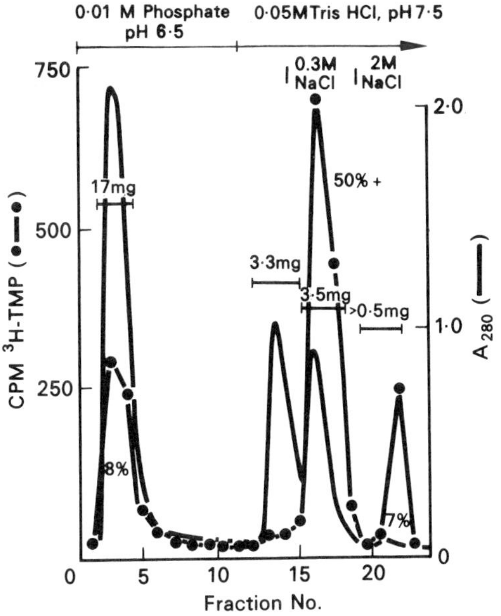

Fig. 3. Attempted purification of DNA polymerase activity of soluble fraction of rat liver. *A 24 mg (22 units) sample of a pH 5.0 fraction (see [17] for details) was loaded onto a rat liver DNA-cellulose column (1.3 x 8 cm) and run as shown.*

Table 1. DNA-cellulose binding of nuclear (N) enzyme and soluble (S1,S2) enzymes. *Values are percentages of loaded material. [Cf. ref. 17.]*

Experiment No.	1	2	3	4	5	6
Enzyme used	S1	S1	S1	N	N	S2
pH	7.5	7.5	6.5	7.5	9.0	7.5
Ionic strength						
0.048	-	26*	8*	-	-	-
0.20	73*	30	21	0*	0*	0*
0.65	12+	14+	43+	25	90	20
2.05	0	0	0	65	12	80
Recovery	85+	70+	72+	90	102	100

The asterisk (*) indicates the ionic strength at which a particular enzyme was loaded on to the same column of rat liver DNA-cellulose prepared as described under Methods. *After loading and standing for 15 min, the column was washed with 20-25 ml of the same buffer followed by similar volumes of buffers of increasing ionic strength as shown. For experiments 1, 2, 4 and 6, buffer A (I 0.048) was used; for experiment 3, 0.00286 M sodium phosphate - 1 mM EDTA, pH 6.5 (I, 0.048); for expt. 5, 0.05 M 2-amino-2-methylpropane-1,3-diol (Ammediol)-HCl containing 1mM EDTA, pH 9.0 (I, 0.03), was used.*

In each case ionic strengths were adjusted to the stated values using NaCl. All buffers and enzymes loaded contained 1 mM dithiothreitol and 0.5 mg/ml bovine serum albumin. The latter did not influence binding of polymerase. Although assays were carried out in ∿20 mM NaCl in expts. 1-3, recoveries are low due to enzyme instability at higher salt concentrations. Amounts of enzyme loaded in expts. 1-6 were 11.0, 11.0, 10.0, 15.3, 13.5 and 12.0 units respectively.

References

1. Alberts, B.M., *Fed. Proc. 29* (1970) 1154.
2. Alberts, B.M., and Herrick, G., *Methods in Enzymology 21* (1971) 198.
3. Cavalieri, L.F., and Carroll, E., *Proc. Nat. Acad. Sci. (Wash.) 67* (1970) 807.
4. Cuatrecasas, P. and Anfinsen, C.B., *Ann. Rev. Biochem. 40* (1971) 259.
5. Poonian, M.S. *et al., Biochemistry 10* (1971) 424.
6. Cuatrecasas, P. *et al., Proc. Nat. Acad. Sci. (Wash.) 61* (1968) 636.
7. Cuatrecasas, P., *J. Biol. Chem. 245* (1970) 3059.
8. Gilham, P.T., *Methods in Enzymology 21* (1971) 191.
9. Gilham, P.T. and Robinson, W.E., *J. Am. Chem. Soc. 86* (1964) 4985.
10. Bautz, E.F.K. and Hall, B.D., *Proc. Nat. Acad. Sci. (Wash.) 48* (1962) 400.
11. Inagoki, A. and Kageyama, M., *J. Biochem. (Tokyo) 68* (1970) 187.
12. Crippa, M., *Nat. (Lond.) 227* (1970) 1140.
13. Litman, R.M., *J. Biol. Chem. 243* (1968) 6222.
14. Knippers, R., *Nat. (Lond.) 228* (1970) 1050.
15. Chang, L.M.S. and Bollum, F.J., *J. Biol. Chem. 246* (1971) 909.
16. Haines, M.E., Johnston, I.R. and Mathias, A.P., *FEBS Lett. 17* (1970) 113.
17. Haines, M.E., Holmes, A.M. and Johnston, I.R., *FEBS Lett. 17* (1971) 63.
18. Meyer, R.R. and Simpson, M.V., *Proc. Nat. Acad. Sci. (Wash.) 61* (1968) 130.
19. Albertsson, P.-Å., *Partition of Cell Particles and Macromolecules*, Wiley, New York (1960).
20. Wagner, A.F., Bugianesi, R.L. and Shen, T.Y., *Biochem. Biophys. Res. Commun. 45* (1971) 184.

12 AFFINITY CHROMATOGRAPHY: ENZYME-INHIBITOR SYSTEMS

K.G. Huggins
Insolubilized Biochemical Laboratory
Miles-Seravac
Maidenhead, U.K.

The classical methods for the isolation of enzymes, such as salt fractionation, pH change and adsorption onto materials such as calcium phosphate, exploit the physical nature of the desired protein. These methods are non-specific and it is necessary to employ a series of these procedures in a multistage purification process. Each stage represents a reduction in final yield and a potential hazard to the enzyme's stability. Methods based on ionic charge forces with ion-exchange materials such as CM- or DEAE-cellulose are also non-specific, relying upon differences in the protein-exchanger charge interactions. Cases are reported where very strong interactions allow good separations, but these are not the general rule when working with crude protein extracts.

A material which allows the simple handling techniques of column chromatography or batch adsorption yet is specific for a desired protein must be of special advantage. The study of enzymes soon revealed that these proteins show a marked specifity towards their substrates and inhibitors. It is logical, therefore, to combine this property with solid state separation methods by chemically binding the inhibitor to a solid support, as with ion-exchangers. As will be indicated, the design and application of such systems is full of technical 'pitfalls' and only recently has the technique become accepted mainly due to the many publications by P. Cuatrecasas [1-4]. Larman in 1953 [5] was probably the first to investigate and succeed with this technique. The use of immunoadsorbents had already been reported; but these systems had proved easier to handle than the enzyme systems. Larman covalently bound the tyrosinase inhibitor p-phenylazophenol to CM-cellulose and, with this packed column, obtained a 60 to 100-fold purification of the enzyme by one passage of crude extract. The subject appears to have been dormant for some time until McCormick in the early 1960's considered the design of affinity media, basing his work upon considerations of the physicochemical factors involved. It is a brief and qualitative consideration of these factors that follows.

Although a highly porous supporting material is desirable, all supports show steric exclusion effects towards macromolecules such as proteins. It is usual for the inhibitor or substrate employed to be smaller than the enzyme protein being isolated. The system must therefore be designed so that the correct area of the inhibitor molecule is exposed to the active site on the enzyme, thus allowing the formation of an enzyme-inhibitor complex. 'Arms', 'spacers' or 'dog-chains' have been employed to maintain the inhibitor between 10 and 24Å from the support backbone. Such techniques have had notable success (6). The common arrangement is for the support and inhibitor to be joined via a C_6 or C_8 carbon chain, for example forming a bridge by reacting support and inhibitor with different amino groups of hexamethylenediamine.

Muscle glycerol-3-phosphate dehydrogenase is not retained well by a column containing glycerol-3-phosphate bound directly to agarose, whereas a bridge of hexamethylenediamine linking to either 1-chloro- or 1-bromo-glycerol-3-phosphate derivatives has furnished efficient affinity media (7) Likewise tyrosine aminotransferase from a tissue culture of mouse hepatoma cells did not bind to an agarose-pyridoxamine phosphate column, but by the introduction of a long-chain bridge a strongly binding and specific column was prepared (2). Although the use of 'spacers' does help to reduce support exclusion effects, there is a much more basic consideration that is often overlooked. This is the bulk of an inhibitor (8).

The bulk tolerance may be considered as the area on the inhibitor molecule not in contact with the enzyme surface in the enzyme-inhibitor complex. This area is important in the design of affinity media but its determination involves laborious organic chemistry. This involves preparing derivatives of the inhibitor, placing bulky groups at different positions on the molecule and observing the effect on the inhibitor kinetics. This bulk tolerance area represents the best position for attachment of inhibitor to the support. It is usually necessary to place suitable reactive groups in this area for this attachment process and here we see the 'pit-falls' noted earlier.

(I)

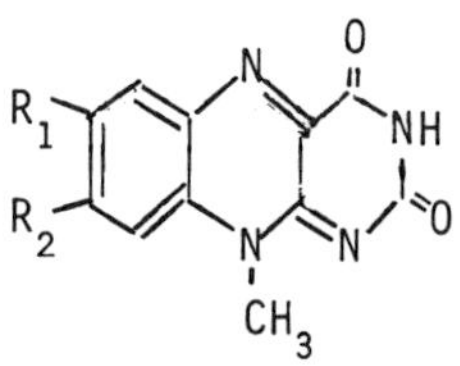

McCormick (9) investigated the attachment of lumiflavin to a support and studied its bulk tolerance towards flavokinase. Derivatives of lumiflavin were prepared by attachment to CM-cellulose of the '$R_2 = NH_2$' compound (I). Marked purification of flavokinase resulted from one passage of extract through a column (100-fold).

Also studied has been a very strong inhibitor of avidin (egg white), biotin, of $K_i = 10^{-15}$ M (II). This compound conveniently has a reactive group R. in the bulk tolerance area. It was bound to cellulose by reaction in pyridine media, and the resulting product showed no ion-exchange properties. Purification of avidin was successfully undertaken (10).

(II)

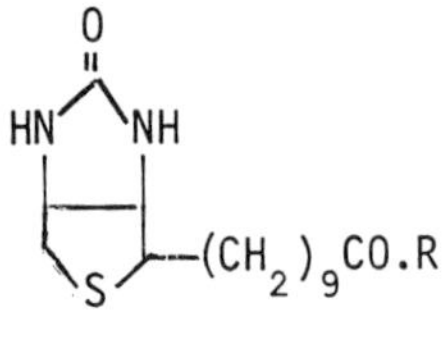

The known dissociation constants of the enzyme-inhibitor complexes (represented by K_i) fall into a wide range.

(Table 1). In designing a suitable column or batch affinity medium, the dissociation constant of the complex is important although it is often necessary to make the best of the range of inhibitors available for a given enzyme. The enzyme must bind to the inhibitor so that separation of the enzyme protein from the bulk solution takes place, but not so strongly that it may not be released for extraction. The K_i values for the inhibitors successfully applied to date fall in a very wide range (10^{-3} to 10^{-15} M) but a workable average may be taken as 10^{-6} M. It is generally reckoned that K_i values greater than 10^{-5} M represent weakly binding enzyme-inhibitor complexes. In such cases either a limited separation must be accepted or special procedures must be employed. The protein could be just retarded within a column and be collected in later elution fractions, or 'spacers' could be used. Substrate columns in place of inhibitor columns can be used for the isolation of enzymes. Of particular interest here are the multi-substrate enzyme systems. Work has been done with NAD^+ and $NADP^+$ bound to solid-phase supports for the isolation of dehydrogenases (11,12).

Attention must now be directed to how these affinity materials can be used. Are they potentially suitable as large-scale isolation media? Their present form is not suited to large-scale application except for some minor enzyme isolations that furnish only small quantities. The next generation of these materials has a potential and future. In Barman's two volumes (13) are detailed some 800 enzymes. It is rather horrifying to consider designing and preparing an affinity material for each enzyme. A column or batch material showing a broad specificity towards a group of enzymes rather than individual enzymes is more attractive. Here the work using NAD^+ columns represent an early form of the new generation affinity materials. The concept may be extended to FAD-dependent enzymes. Metal radicles such as organomercury bound to solid supports can be used for

Table 1. K_i Values for enzyme-inhibitor systems (2).

ENZYME	INHIBITOR	K_i, *M*
Avidin	biotin	10^{-15}
Staphylococcal nuclease	3'(4-aminophenyl-phosphoryl)-deoxythymidine-5'-phosphate	10^{-6}
Chymotrypsin	N-acetyl-D-tryptophan ester	10^{-4}
β-Galactosidase	p-aminophenyl β-D-thiogalactopyranoside	5×10^{-3}
Glycerol-3-phosphate dehydrogenase (mitochondrial)	1-chloroglycerol 3-phosphate	2×10^{-3}
Acetylcholinesterase	ε-aminocaproyl-p-aminophenyl-trimethyl ammonium bromide	6×10^{-6}

the isolation of sulphydryl enzymes. Already papain (14) and bromelain (15) have been extracted using such column material.

This is a young technique which is proving to be very useful in enzyme isolation, certainly on the laboratory scale. With sound physico-chemical considerations as the basis, it should be possible to develop from the laboratory a technique of wide applicability in both laboratory and industry, finding its place with the classical methods for enzyme isolation and production.

References

1. Cuatrecasas, P. and Anfinsen, C.B., *Methods in Enzymology 22* (1971) 345.
2. Cuatrecasas, P. and Anfinsen, C.B., *Ann. Rev. Biochem. 40* (1971) 259.
3. Jerina, D.M. and Cuatrecasas, P., *Proc. 4th Int. Congr. Pharmacol., 1969;* Schwabe, Basel (1970).
4. Cuatrecasas, P., in *Biochemical Aspects of Solid State Chemistry* (ed. G.R. Starke), Academic Press, New York (1971).
5. Lerman, L.S., *Proc. Nat. Acad. Sci. (Wash.) 39* (1953) 232.
6. Cuatrecasas, P., *Nat. (Lond.) 228* (1970) 1327.
7. Halohan, P.D., Mahajan, K, and Fondy, T.P., *Fed. Proc. 29* (1970) 888.
8. Barker, B.R., *Design of Active-Site-Directed Irreversible Enzyme Inhibitors,* Wiley, New York (1967).
9. McCormick, D.B., Arsenis, C. and Hemmerick, P., *J. Biol. Chem. 238* (1963) 3095.
10. McCormick, D.B., *Anal. Biochem. 13* (1965) 194.
11. Lowe, C.R. and Dean, P.D.G., *FEBS Letters 14* (1971) 313.
12. Larsson, P.O. & Mosbach, K., *Biotech. Bioeng. 13* (1971) 393.
13. Barman, T.E., *Enzyme Handbook,* Springer-Verlag, Berlin (1969).
14. Slurterman, L.A. and Wijdenes, J., *Biochim. Biophys. Acta 200* (1970) 593.
15. Murachi, T., *Methods in Enzymology 19* (1970) 277.

13 PROTEIN PURIFICATION BY IMMUNOADSORPTION

Oliver Cromwell
Wolfson Bioanalytical Centre
University of Surrey
Guildford, Surrey, U.K.
and
Wellcome Research Laboratories
Langley Court
Beckenham
Kent BR3 3BS, U.K.

The problem of purifying specific proteins from heterogeneous mixtures, such as serum, is usually overcome by employing physico-chemical techniques such as electrophoresis, selective precipitation, ion-exchange chromatography or gel filtration. A high degree of purity can usually be obtained only through several precipitation steps, and hence the final yields are small. The idea of using protein insolubilized by covalent bonding to an inert carrier for the purification of specific antibody was applied by Campbell, Luescher and Lerman [1] when they used bovine serum albumin (BSA) linked to p-aminobenzylcellulose to remove the specific antibody from rabbit antiserum. Subsequently the potential of so-called immunoadsorption for both antibody and antigen purification has slowly come to be realized. This article fills in the background: it largely comprises a survey, not aimed at the specialist, of the preparation and relative merits of different immunoadsorbents.

Immunoadsorption requires either an antiserum showing specificity for the particular immunogenic protein to be purified or, if antibody purification is to be attempted, pure antigen.* The protein is first insolubilized by linking it to a suitable matrix, and any protein which does not bind is washed off. Unreacted sites on the matrix are blocked by reaction with amino acids or other suitable compounds. The immunoadsorbent thus prepared is suspended in the biological fluid containing the protein of interest, or else the fluid is passed through a column packed with the adsorbent. Specific antibody-antigen binding then occurs, while all non-specific proteins remain free to be washed from the adsorbent by a suitable buffer. The immunochemically linked protein is then eluted using a solution of suitable pH and/or ionic strength.

Although appearing to be a powerful technique, immunoadsorption presents many problems. The first is to obtain a suitable specific antiserum, which may take several months, or a very pure sample of the antigen. In

* *The distinction between 'antibody' and 'antigen' is largely operational: an antibody from one species may be immunogenic if injected into another.*-Ed.

the case of the antiserum, the antibodies should ideally have a good avidity for the antigen, yet the antibody-antigen complex must be readily dissociable. The problems of raising antisera are covered elsewhere (e.g. 2). For insolubilizing the protein there are various procedures (3-7) some of which are covered later. Besides their uses for antigen purification, antibody immunoadsorbents can be used for solid-phase radioimmunoassay, this assay technique being quicker and more straightforward than the tube method or the double antibody method.

Another purification technique making use of an insolubilized ligand to bind the protein is affinity chromatography, which has proved useful in the nucleic acid field and especially for enzyme purifications *[Articles 11 & 12]*. Enzymes as well as enzyme inhibitors have been insolubilized, and are in fact finding applications in industry, particularly in brewing and for the clarification of fruit juices, and in the laboratory for simulating natural systems and for enzyme assays. Enzymes covalently linked to nylon tubing (8) or similar support materials furnish a powerful tool for automating routine assays in clinical biochemistry.

SPECIFIC ANTISERA AND ANTIBODIES

The following notes may help readers unfamiliar with immunology, in relation to the above-mentioned requirements that the antiserum should be monospecific, have a high avidity and furnish antibody-antigen complexes that are easily dissociable. Proteins of molecular weight >5,000 are usually good immunogens, but with smaller molecules it may be necessary to link them to a carrier such as bovine serum albumin before using them to immunize an animal. The best results are obtained when the injection is given in Freund's complete adjuvant. The antiserum obtained from an animal very often shows cross-reactions to other proteins; hence it is common practice to immunize with a crude preparation of the protein rather than a highly purified one which may be difficult to ontain. Furthermore, impurities often have an adjuvant effect. Before using the antiserum to prepare an immunoadsorbent the cross-reactions must be removed. This is achieved either by liquid adsorption or by using an immunoadsorbent prepared from the cross-reacting proteins.

All but the very smallest protein molecules have more than one antigenic determinant, and because of this an antiserum raised against a particular protein will contain a heterogeneous population of antibodies. All antibodies belong to one of the different classes of immunoglobulins - IgG (the commonest, Fig. 1), IgM, IgA, IgD and IgE. Each molecule of IgG consists of two heavy and two light chains, having molecular weights of 56,000 and 23,000 respectively, held together by interchain disulphide bonds, the number of which depends upon the class and subclass of the immunoglobulin. The three-dimensional structure of the molecule is strongly influenced by intra-chain disulphide links. Both light and heavy chains contain regions of variant and invariant amino acid sequence. In light chains the 106 amino acids from the C-terminal end are invariant, whilst the remainder to the N-terminal end show variations in sequence and composition. A variant region also occurs within the Fd region of the heavy chains. Differences in amino acid composition in those variant regions allow for antibody specificity.

There are two antibody active sites per IgG molecule, one associated with the N-terminal ends of each pair of light and heavy chains. The F_c region is not involved in the active site. For an antibody to retain its antigen-binding ability, once it is in an insolubilized preparation, the active sites must be clear of the carrier and not suffer steric hindrance.

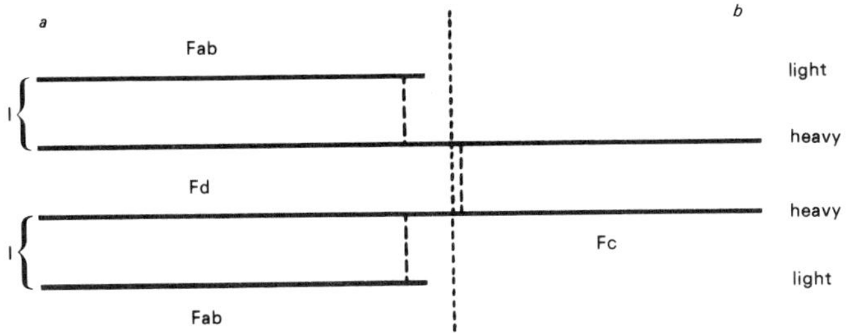

Fig. 1. Structure of the IgG molecule, with heavy chains (μ, α or γ) and light chains (κ or λ). *a* = N-terminal ends, *b* = C-terminal ends, I = antigen-binding sites.

PREPARATION OF IMMUNOADSORBENTS

Considerations in the choice of method

The methods available for either coupling proteins to insoluble supports or polymerizing them are numerous and diverse, reflecting the fact that no one technique is generally applicable to all systems. Some of the preparations necessarily take several days and need a number of different steps, whilst others can take as little as a day to complete. One can only ascertain which system is most suited to the job in hand by testing each one, judging it by the criteria outlined here. A simplified scheme for preparation of immunoadsorbents is given in Fig. 2.

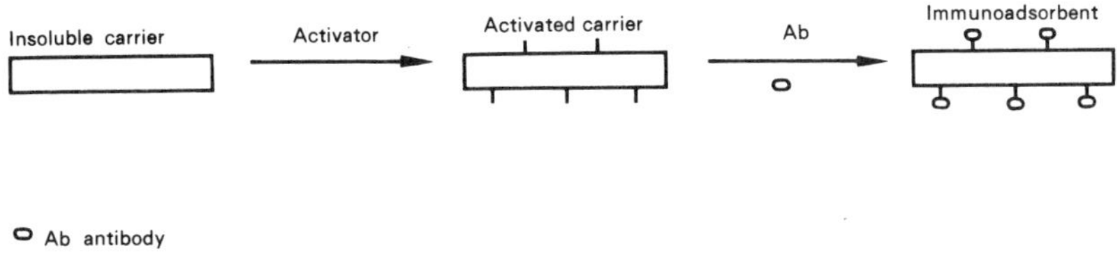

Fig. 2. Preparation of immunoadsorbents. *The activation step may involve more than one reaction.*

Methods involving physical adsorption of the protein onto carriers such as activated charcoal, erythrocytes, latex particles, kaolin and ion-exchange materials, are unsatisfactory because it will almost certainly be the case that one will encounter conditions of pH and ionic strength which will dissociate the protein from the carrier. Preparations involving trapping of the antibody molecules within a lattice structure, such as that formed by polyacrylamide, also have their shortcomings. Provided that the antibody molecules cannot escape through the pores of the lattice work, and that the antigen is small enough to enter, the immunoadsorbents should perform satisfactorily. These problems were overcome by Carrel and Barandun (9) when they used a macroporous polyacrylamide gel to prepare a number of immunoadsorbents of high capacity, and by Carrel *et al.* (10) in isolating normal immunoglobulin E.

Preparations involving covalent linking of the protein are preferable to those mentioned above, since it is possible to choose the type of bond which will be stable under the conditions likely to be encountered by the adsorbent. Furthermore, one can control which type of amino acid residues a: are involved in the links. Protein polymers are easily made by reaction of protein with cross-linking reagents such as ethyl chloroformate (11), glutaraldehyde (12) or bis-diazotized benzidine (13). Disulphide-linked antibodies have been prepared successfully (14,15) by thiolating the protein with N-acetylhomocysteine thiolactone and then cross-linking with a potassium ferricyanide reagent or else with bifunctional organomercurial linking agents (16). Versatile immunoadsorbents have been prepared by conjugating proteins with ethylene-maleic anhydride (17). Covalent coupling to insoluble supports such as cellulose, beaded agarose, Sephadex, beaded polyacrylamide, polystyrene, nylon and glass is the most commonly used method of insolubilization. This is discussed in more detail in the next Section.

The following criteria are useful when assessing the relative merits of immunoadsorbents:- (1) the method of preparation must produce insolubilized protein which will not dissociate from the carrier under the conditions of adsorption and elution, and which will remain stable on storage; (2) the conditions of coupling must be such that neither the protein to be coupled nor the matrix are denatured; (3) the insolubilized protein must retain a high proportion of its original specific biological activity; (4) the preparation should result in insolubilization of a high percentage of the added protein, provided of course that the protein remains biologically active; (5) the adsorbent should be suitable for either batchwise or column use; (6) elution must result in recovery of a high percentage of specific protein taken up by the immunoadsorbent, and the protein must still be biologically active; and (7) the adsorbent should be capable of re-use.

Conditions (1) can be fulfilled by using covalent bonds to effect insolubilization. Once the reaction is complete the preparation must be washed exhaustively to ensure that all uncoupled protein is removed. This may take several days and must involve washings with all buffers and eluents to which the immunoadsorbent is to be exposed. The efficiency of the washing procedure is best monitored by including a small percentage of radioactively labelled protein in the reaction mixture. Alternatively the washings can be concentrated by a known factor and the protein determined spectrophotometrically. When washing is complete the amount of coupled protein may be determined by counting bound radioactivity or estimating protein after acid or alkaline hydrolysis. With polysaccharide-based immunoadsorbents it may be possible to release the protein by enzymic digestion of the carrier.

Most preparations can be carried out in aqueous buffers within the pH range 6-9, but some methods, such as glutaraldehyde polymerisation, call for a slightly lower pH, and a few reagents have to be dissolved up in organic solvents such as dicyclohexylcarbodiimide.

The amount of biological activity retained by insolubilized antibody can easily be assessed by radioimmunoassay procedures. If suspensions of the adsorbents are first diluted to the same antibody concentration a direct comparison can be made. Alternatively the assay may be carried out using dilutions of the immunoadsorbents. Both methods have been used successfully in these laboratories.

If one of the two antibody-binding sites in IgG is to bind an antigen, it must not be blocked by either covalent linkages in the carrier or the close proximity of the carrier. Although it is possible to choose a method of binding which involves a particular amino acid, one cannot dictate which residues are available to react, and it is inevitable that some covalent links will be formed in or near the active site. The chances of this happening can be reduced by (i) employing the minimum number of covalent linkages, and (ii) locating the antibody as far from the surface of the matrix as possible (Fig. 3). Evidence concerning (i) in relation to antibody activity has been obtained by use of cellulose derivatives with

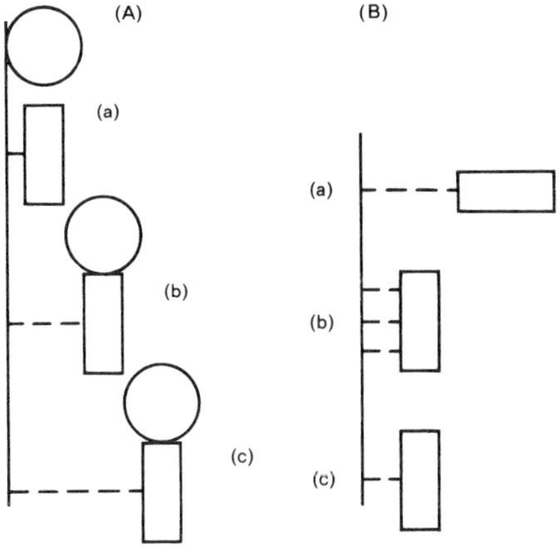

Fig. 3. Types of protein linkage to an insoluble carrier.

A: Proteins linked to the surface of the carrier via aliphatic chains of increasing length.
(a) Inactivated by close proximity of the carrier;
(b) & (c) active.

B: Proteins linked to the surface via -
(a) a single covalent bond,
(b) several covalent bonds,
(c) a single covalent bond.

varying degrees of substitution. Concerning (ii), Cuatrecasas (18) prepare insolubilized ligand by attachment to Sepharose either directly (none bound or via aliphatic chains of different lengths. With a 2-C chain derived from ethylene diamine and with the tripeptide Gly-Gly-Tyr inserted between the carrier and the ligand, increasing amounts of enzyme were bound. His ligands being much smaller than antibodies, one would not expect antibody activity to be increased as dramatically by insertion of a short aliphatic chain. Work in our laboratories has shown that antibody activity is increased by linkage through longer aliphatic chains, but there is a suggestion that these may fold back on themselves, so bringing the antibody closer to the carrier. With a larger number of linkages it is reasonable to expect that the structure would have more rigidity, but then the chances of blocking the active site are increased. Antibodies having only two antigen-binding sites are much more likely to be inactivated by insolubilisation than are antigens which have several antigenic determinants showing specificity for different antibodies.

The immunoadsorbent has to be stable, particularly if it is to be used repeatedly. Cellulose-based immunoadsorbents retain activity well afte 3 months' storage at 4° in a protein-containing buffer, and likewise after lyophilisation. Hydrophobic carriers such as polystyrene may tend to denature the antibody, while hydrophilic carriers may stabilize the protein configuration and are therefore to be preferred. Strongly anionic or cationic carriers produce electrostatic interactions with the covalently linked antibody and so influence its stability and activity. The pH optimum of the antibody-antigen reaction and its sensitivity to temperature and ionic strength may well be influenced by the carrier. The carrier's surface area will also affect antibody activity (19).

Whether the immunoadsorbent is to be used in a column or batchwise rather depends on the source from which the antigen is to be purified and the method of elution to be used. If the volume is small a column procedure will be satisfactory; the sample running, washing and elution can be done reasonably quickly. If a batch technique is to be used to extract antigen from a small volume, it is important that the immunoadsorbent have a high antibody content in relation to the adsorbent volume. With large volumes the batch procedure is much quicker, the adsorbent being suspended in the solution and then centrifuged out. Washing is done by repeated re-suspension in a suitable buffer prior to elution. For column work, adsorbents prepared using beaded agarose or polyacrylamide, Sephadex, or glass beads are best since they pack well to give uniform beds with good flow rates. Cellulose and protein polymer particles tend to be irregular in shape and, as a result, columns packed with these materials have poor flow rates. They are, however, more suitable than the beaded carriers for batchwise adsorbtions since they settle out of suspension much more slowly.

After preparing a particular immunoadsorbent the unreacted active groups of the carrier must be blocked off so that they will not react with protein added subsequently. This may be done with glycine where the covalent links involve amino groups, cysteine in the case of thiol groups, and other suitable amino acids. Aldehyde groups are blocked with sodium borohydride or bisulphite, and diazonium groups with β-naphthol. With cyanogen bromide-activated carriers and bromoacetyl derivatives, ethanolamine effectively neutralizes the active groups and also removes any ion-exchange properties imparted to the carrier by those groups.

Some methods of preparation

Of the various linking methods as described in reviews (5, 7) and papers (e.g. 18), only some common ones will be touched on. The functional groups available in proteins to form covalent bonds are N-terminal amino, ε-amino, C-terminal carboxy, β- and γ-carboxy, serine -OH, the tyrosine phenolic ring, and -SH, indole, imidazole and guanidine.

The diazo group was the first used to bind protein to cellulose by Campbell *et al.* (1) and subsequently by Gurvich *et al.* (19). They used *p*-aminobenzylcellulose and *m*-aminobenzyloxymethylcellulose respectively, prepared by etherification reactions of cellulose and then reacted with protein, primarily via amino and imidazole groups. Phenolic, indole and guanidine groups may also be involved. Diazonium derivatives have the advantage that they can react with proteins over quite varied conditions, and comparable coupling levels are obtained over the pH range 6-9. Diazotized polyaminostyrene has been used to insolubilize antigens (20), but the variability of the physical properties of the derivative make it unpredictable and difficult to work with.

Antibodies to human IgG have been successfully coupled to carboxymethylcellulose in our laboratories using the acyl azide derivative (cf. 21), but the product invariably proved to be difficult to work with. The reaction with amino, hydroxyl and thiol groups is by nucleophilic substitution, producing amide, ester and thioester bonds respectively. With bromoacetylcellulose (22) the covalent coupling is primarily through amino groups at pH 9.0, whereas the imidazole group of histidine is involved at pH 7.0.

Immunoadsorbents have been successfully prepared in the Beckenham laboratories with 2-amino-4,6-dichloro-s-triazine, which with chymotrypsin gave better derivatives than other triazine compounds (23), and has worked with cellulose, carboxymethylcellulose, DEAE-cellulose and Sepharose. Antigens have been coupled to carboxymethylcellulose in the presence of N,N'-dicyclohexylcarbodiimide (5), the reaction being favoured by acid pH; yet it goes satisfactorily at pH 6.0, making it unnecessary to expose the antigen to low pH (24). The water-soluble carbodiimides seem not to work well, despite favourable reports.

Oxidation of cellulose with periodate converts vicinal hydroxyl groups to aldehydes, and has enabled bovine serum albumin to be insolubilized (25). Isocyanate derivatives of cellulose, dextran and Sephadex conjugate with peptides and proteins, likewise via amino groups (26). For carbohydrates such as cellulose and Sephadex, cyanogen bromide has become the most popular reagent, giving rise to imidocarbonic acid esters and carbamate groups which react with protein (27); the imidocarbonate group is not formed in Sepharose because there are no vicinal hydroxyls. The gentleness of the coupling is helpful for immunoadsorbent preparation. The degree of activation is pH-dependent, being maximal at pH 11 or above. Cyanogen bromide-activated polysaccharides have been used for preparing various derivatives (18); thus, ω-aminoalkyl groups on the carrier may be reacted with protein amino groups (in the presence of glutaraldehyde) or carboxyl groups (activated by N,N'-dicyclohexylcarbodiimide), or may be used to prepare, e.g., bromoacetyl and -SH derivatives.

Beaded polyacrylamide gels have a great potential for the preparation of immunoadsorbents, their carboxamide groups being easily modified (28). Aminoethyl and hydrazide derivatives serve as parent compounds for a range of derivatives, as with cyanogen bromide-activated polysaccharides. Hydrophilic co-polymers based on polyacrylamide are marketed as 'Enzacryl', with variants allowing coupling to aromatic residues and amino groups in proteins.

With glutaraldehyde, giving another widely used type of immunoadsorbent, the polymers can be made either by direct polymerisation of the protein or by linking the protein to a carrier polymer made by cross-linking bovine serum albumin with glutaraldehyde (12). Different proteins show different pH optima for polymerisation with glutaraldehyde and ethyl chloroformate (11), a property associated with their isoelectric points. Polymerization can also be achieved by cross-linking thiolated proteins (14), or by using ethylene maleic anhydride (17) or bis-diazotized benzidine (13).

Efficiency of protein linking and retention of activity

Immunoadsorbents obtained by polymerisation techniques using bifunctional reagents such as glutaraldehyde and by trapping the protein inside a lattice of, say, polyacrylamide give preparations containing better than 90% of the added protein. Ethyl chloroformate has given more than 95% polymerisation of goat antiserum (29), and glutaraldehyde has given 100% with gamma globulin and bovine serum albumin at the optimal pH (12). However, the very nature of these techniques is wasteful of active protein in that some is sacrificed in the polymerisation reaction or masked by the adsorbent's structure. With a correction for non-specific adsorption, 1 mg of antibody to human gamma-globulin in an ethyl chloroformate polymer could bind 0.2 mg of the antigen (29). Since these figures were close to the molecular ratio of human gamma-globulin to specific antibody in the equivalence zone of the precipitin reaction, antibody activity appeared to be little impaired by the polymerisation. However, with a glutaraldehyde polymer, it was reported (12) that 4-5 times more insolubilized than soluble normal human serum was necessary to adsorb the same quantity of antibodies. The capacity of the adsorbent was little affected by raising the glutaraldehyde concentration, suggesting that only certain amino groups were available for cross-linking; moreover, the adsorption capacity of the polymer depended on the nature of the antigen used. With ethyl chloroformate, unlike glutaraldehyde, the activity varied markedly with the polymerisation pH (11).

In covalently linking protein to cellulose or other carriers the amount bound depends strongly on the method, the availability and activity of the carrier binding sites being all-important. Bromoacetyl cellulose may bind over 90% of the added protein up to the capacity of the carrier (22). With other preparations the percentage of available protein attached increases with the protein concentration in the reaction mixture. The binding capacities of carbohydrate-based immunoadsorbents vary considerably with the protein and the derivative employed. Egg-white lysozyme on bromoacetyl cellulose (22) was 30 times less efficient in binding antibody than was the free antigen. Yet antibody thus purified may be almost completely precipitable. For proteins coupled to cellulose derivatives, 5-8% insolubilisation was been found with N,N'-dicyclohexylcarbodiimide and 20-40% with a water-soluble diimide (24); both products performed well, indicating that a high protein content does not necessarily make for a better adsorbent.

General methods for measuring protein incorporation in immunoadsorbent synthesis include use of isotopically labelled proteins, alkaline digestion of the immunoadsorbent and Lowry determination of the protein, measurement of unbound protein in supernatant and washings, and total amino acid analysis.* To determine immunoadsorbent capacity, one way is to measure the protein content of an aliquot and then use an identical aliquot to adsorb the specific protein; a second protein determination is made and the capacity determined by difference. There must be excess of the pure protein to ensure saturation of the adsorbent and avoid any errors due to non-specific adsorption. Adsorption of the specific protein followed by washing of the immunoadsorbent, elution and protein measurement in the eluate is unsatisfactory because some of the less tightly bound protein will probably be lost during the washing and, moreover, elution itself will not usually result in dissociation of all the specifically bound protein. Such tests can be conducted using column or batch techniques. The best way of measuring immunoadsorbent capacity is by radioimmunoassay, which also allows a parallel experiment to determine the binding capacity of the free protein using a double antibody technique as exemplified by insulin assay. Guinea-pig anti-insulin is incubated with free insulin and an aliquot of ^{125}I-insulin; finally the guinea-pig antibodies with bound insulin are precipitated using a specific anti-guinea-pig-globulin serum. In this method, which is very sensitive, we measure capacity in terms of the protein concentration necessary to cause 50% inhibition of binding of the label.- The larger the capcity of the immunoadsorbent, the greater the protein concentration necessary to cause the 50% inhibition. The technique involves microscale work. The experiment can be done on a macroscale and the protein concentrations determined spectrophotometrically. In such comparative experiments it is essential to run blanks with insolubilized preparations of non-specific antibody or antigen to allow for possible non-specific adsorption.

* *An example of a special method is iron estimation in myoglobin studies [30].*

USE OF IMMUNOADSORBENTS

Practical aspects of immunoadsorption and elution (Fig. 4) are now considered. Immunoadsorption is commonly carried out at room temperature, but occasionally at 4° or, where acceleration is advantageous to the system, at 37°. High avidity antibody will be the first to complex with antigen. Non-specific adsorption is an important problem. An immunoadsorbent will be unsuitable if excess carrier binding sites have not been blocked off or if the carrier has ion-exchange properties effective at the pH and ionic strength of the system. Non-specific adsorption was reported to be negligible with ethyl chloroformate-polymerized protein and also with immunoadsorbents prepared using glutaraldehyde (11). However, experience has shown that when extraction of proteins from whole serum is attempted with glutaraldehyde-polymerized specific antiserum, the eluate invariably contains a protein mixture. This may be due to non-specific interactions between serum proteins and non-antibody protein in the immunoadsorbent, as likewise postulated (17) in a study of pollen linked to ethylene maleic anhydride-polymerized bovine serum albumin: non-specific adsorption to the albumin was thought to account for impurity of the eluted andibodies. Non-specific adsorption, e.g. of serum components (22), is minimal with non-proteinaceous carriers such as cellulose. Adsorption to cellulose at pH 3-4 of proteins which are released by 0.1 N NaOH but not at pH 7 or 1 was thought to be due to hydrogen bonding, insofar as acetylated or ethylated carboxymethylcelluloses desorbed the protein at pH 7, their hydroxyl groups being largely blocked (5). Synthetic polymers such as polyacrylates show high non-specific protein binding.

The binding of antibody and antigen on cellulose immunoadsorbents may become tighter on long standing (1). The time to achieve maximum binding will depend on the avidity of the antiserum. In our experience with several rabbit and sheep antisera, more than 80% of available antibody-binding sites are saturated within 1 hour using batch adsorption procedures

Elution is the most difficult problem. The ideal eluent will alter the protein conformation sufficiently to decrease its affinity for the anti-

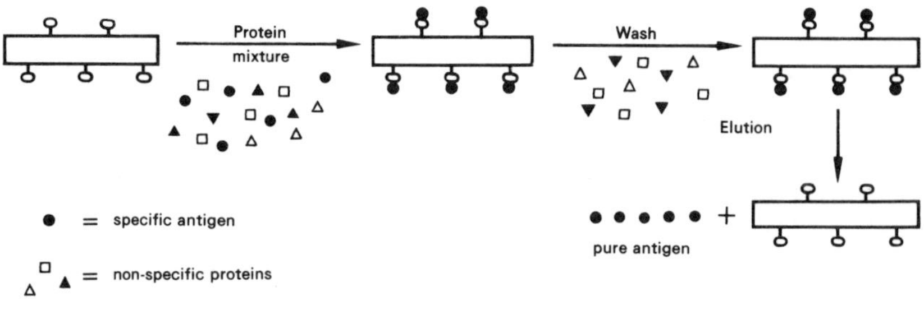

Fig. 4. Immunoadsorption, and elution of the specifically adsorbed protein

body and allow dissociation of the antibody-antigen complex but will not cause denaturation of the protein. Protein three-dimensional structure is maintained by relatively weak bonds (besides disulphide bonds) some of which are broken by urea and guanidine or by chaotropic ions (31) such as thiocyanate, perchlorate and iodide. Where complete dissociation proves very difficult, disulphide bonds may be involved.

With chaotropic ions, too strong a pH or too high an ionic strength had irreversible effects on the antibody (31), although most of it may be recovered under suitable conditions. The removal of chaotropic ions can be achieved only by dialysis; hence the native protein structure will be regained only after some time.

Another method of elution employs low pH as achieved with HCl, glycine-HCl buffers, acetic acid and propionic acid. This has the advantage that the eluent can be immediately neutralized by transfer to a suitable buffer. An alkaline pH, being likewise far from the isoelectric point, may be equally effective. Acid and alkaline pH elutions work either with columns or batchwise, whereas chaotropic ions work effectively only batchwise insofar as they act by solvolysis of the antibody-antigen complex rather than by serving as a dissociating agent.

In a comparison of various immunoadsorbents (30) with insolubilized myoglobin and a rabbit antiserum, 20 mM HCl-0.15 M NaCl (pH 1.8) and 1 M propionic acid (pH 2.5) gave the highest yields of protein, but the latter gave the highest yields of precipitable antibody, e.g. 77% with Sepharose 4B, 55% with carboxymethylcellulose, 58% with aminoethylcellulose, 109% with bromoacetylcellulose, but nil with use of ethylene-maleic anhydride at pH 6 or 8. The latter polymers decreased in volume on treatment with the propionic acid, and it was suggested that the antibody could have been occluded.

Work by Hill (32) has shown that dioxane promotes the release of antibody when mixed with the weak organic acids such as propionic acid: with rabbit anti-human serum albumin antibody, its increased release was accounted for by antibodies of higher affinity than those eluted by the acid alone. The dioxane is thought to encourage hydrophobic bond formation.

With electrolyte solutions of increasing concentration to fractionate an antibody population (33), the association constants were found to tend to zero at high concentrations. With the eluents used the amount of antibody recovered at high ionic strength did not vary significantly (Table 1). In a study with rabbit anti-bovine serum albumin antibodies (17), sequential treatment of the immunoadsorbent with glycine buffers of decreasing pH (3.0, 2.5 and 2.0) gave three distinct fractions, the first being low in hyperimmune animals where the proportion of low affinity antibodies is low. Various factors affecting antibody recovery were studied. With pH 2.8 glycine-HCl, ionic strength 0.2, human serum albumin antibodies were maximally eluted in 15 min at 20^0 but took 90 min at 4^0.

Table 1. Elution by different electrolyte solutions of antibodies fixed on an immunoadsorbent, viz. glutaraldehyde-polymerized protein (from ref. 33).

Human serum albumin antibodies (23 mg) prepared in rabbits were adsorbed on the immunoadsorbent. Elution was performed with 2 x 10 ml of 2.5 M solution followed by 1 x 10 ml of 5 M solution.

ELUTING SOLUTION	pH OF THE SOLUTION	ANTIBODY ISOLATED, mg
NaSCN	7.0	17.5
NH_4I	9.0	17.7
LiI	9.0	18.4
NaI	9.0	19.0
$MgCl_2$	7.0	19.0
Glycine-HCl buffer	2.8	19.7

During work in the Beckenham laboratories with immunoadsorbents prepared by coupling sheep anti-human IgG (specific for F_c region) to various cellulose derivatives, different eluents have been employed (Table 2). Besides factors listed above, the nature of the cellulose derivative, the degree of substitution of active groups on the cellulose, and the level of substituted antibody have been shown to influence the amount of specific antigen eluted. Table 2 shows recovery values for two adsorbents (22) with different degrees of substitution of the *m*-aminobenzyloxymethyl group as judged by N content. The more substituted adsorbent gave slightly better % recoveries with most eluents, and markedly better recoveries (the yield being notably high) with acetic or propionic acid. Bromoacetylcellulose gave similar results (not tabulated), acetic and propionic acids again being the most efficient with the adsorbent having the fewest covalent links.

Table 2. Recovery of human IgG with different eluents and with differing degrees of substitution of the cellulose adsorbent.

Some of the eluents were as used in a reported study [30]. The % recovery values, determined using trace-labelled antigen, are means from 4 experiments, agreeing within ±2%. Adsorbent 1 had four times as many available diazo groups as Adsorbent 2 (based on N content).

ELUENT	% RECOVERY	
	adsorbent 1	adsorbent 4
3 M sodium thiocyanate, 50 mM phosphate, pH 6	77	83
3 M sodium thiocyanate, 50 mM Tris-HCl, pH 9	75	77
50 mM glycine-HCl, pH 2.8	75	82
1 M acetic acid, pH 2.3	60	88
1 M propionic acid, pH 2.5	58	86
8 M urea - 10 mM phosphate - 150 mM sodium chloride, pH 7.6	70	76
2.5 M magnesium chloride	46	48

References

1. Campbell, D.H., Luescher, E. & Lerman, L.S., *Proc. Nat. Acad. Sci. (Wash.) 37* (1951) 575.

2. Hurn, B.A.L. & London, J., in *Radioimmunoassay Methods* (K.E. Kirkham & W.M. Hunter, eds.), Livingstone, Edinburgh (1971) p. 63.

3. Kay, G., *Process Biochemistry 3(8)* (1968) 36.

4. Silman, I.H. & Katchalski, E., *Ann. Rev. Biochem. 35* (1966) 873.

5. Weliky, N. & Weetall, H.H., *Immunochemistry 2* (1965) 293.

6. Sehon, A.H., in *International Symposium on Immunological Methods of Biological Standardization (Royaumont, 1965) - Symp. Series Immunobiol. Standard. 4,* Karger, Basel (1967), p. 51.

7. Campbell, D.H.& Weliky, N., in *Methods in Immunology and Immunochemistry,* Vol. 1 (C.A. Williams & M.W. Chase, eds.), Academic Press, New York (1967), p. 365.

8. Hornby, W.E. & Filippuson, H., *Biochim. Biophys. Acta 220* (1970) 343.

9. Carrel, S. & Barandun, S., *Immunochemistry 8* (1971) 39.

10. Carrel, S., Theilkäs, L., Morell, A., Skvaril, E. & Barandun, S., *Biochem. J. 122* (1971) 405.

11. Avrameas, S. & Terynck, T., *J. Biol. Chem. 242* (1967) 242.

12. Avrameas, S. & Terynck, T., *Immunochemistry 6* (1969) 53.

13. Ishizaka, K. & Isizaka, T., *J. Immunol. 93* (1964) 59.

14. Stephen, J., Gallop, R.G.C. & Smith, H., *Biochem. J. 101* (1966) 717.

15. Crook, N.E., Chidlow, J.W., Stephen, J. & Smith, H., *Immunochemistry 9* (1972) 749.

16. Mandy, W.J., Rivers, M.M. & Nisonhoff, A., *J. Biol. Chem. 236* (1961) 3221.

17. Centeno, E.R. & Sehon, A.H., *Immunochemistry 8* (1971) 887.

18. Cuatrecasas, P., *J. Biol. Chem. 245* (1970) 3059.

19. Gurvich, A.E., Kuzovleva, O.B. & Tumanova, A.E., *Biokhimiya 26* (1961) 934.

20. Sehon, A.H., *Brit. Med. Bull. 19* (1963) 183.

21. Mitz, M.A. & Summaria, L.J., *Nat. (Lond.) 189* (1961) 576.

22. Robbins, J.B., Haimovich, J. & Sela, M., *Immunochemistry 4* (1967) 11.

23. Kay, G. & Lilly, M.D., *Biochim. Biophys. Acta 198* (1970) 276.

24. Dandliker, W.B., de Saussure, V.A. & Levandoski, N., *Immunochemistry 5* (1968) 357.

25. Sanderson, C.J. & Wilson, D.V., *Immunology 20* (1971) 1061.

26. Axén, R. & Porath, J., *Nat. (Lond.) 210* (1966) 367.

27. Axén, R. & Ernback, S., *Eur. J. Biochem. 18* (1971) 351.

28. Inman, J.K. & Dintzis, H.M., *Biochemistry 10* (1969) 4074.

29. Radermecker, M. & Goodfriend, L., *Immunochemistry 6* (1969) 484.

30. Boegman, R.J. & Crumpton, M.J., *Biochem. J. 120* (1970) 373.

31. Dandliker, W.B., Alonso, R., de Saussure, V.A., Kierszbaum, F., Levison, S.A. & Schapiro, H.C., *Biochemistry 6* (1967) 1460.

32. Hill, R.J., *J. Immunological Methods 1* (1972) 231.

33. Terynck, T. & Avrameas, S., *Biochem. J. 125* (1971) 297.

14 ASPECTS OF THE USE OF PHENOL IN THE ISOLATION OF RNA*

R. Williamson
Beatson Institute for Cancer Research
Hill Street
Glasgow, C.3.
Scotland

The following comments concern the use of denaturing agents in the preparation of nucleic acids. Few results are given; the aim is rather to draw attention to some of the questions which are still outstanding.

Unlike Dr. Pusztai, whose paper *[Article 16]* concerns the use of phenol to isolate proteins, nucleic acid biochemists rely on phenol to completely denature proteins. Such is the currency of this procedure that we rarely question the effect phenol may have upon the product. I now consider whether this is justified in the case of messenger RNA (mRNA).

The isolate of nucleic acids with phenol is, of course, linked with the name of the late Dr. Kirby of the Chester Beatty Research Institute who reported the technique in 1956 (1) and continued to improve it until his death in 1967. Cresol is added with phenol to improve deproteinisation (2) and detergents such as sodium dodecyl sulphate, sodium dodecyl sarcosine, napthalene disulphonate or tri-isopropylnapthalene-sulphonate are added not only to improve dissociation of protein from nucleic acid but also to help inhibit nucleases (3).

This method, when applied to mouse or rabbit reticulocyte polysomes, gives preparations of mRNA after zonal ultracentrifugation which are active in directing globin synthesis in heterologous cell-free systems. These preparations, however, show considerable secondary structure and may migrate anomalously in polyacrylamide gels (4). Moreover, in high-concentration analytical polyacrylamide gels, they show a band pattern which cannot be immediately correlated with our preconceptions on globin mRNA species and *in vitro* messenger activity (5). We considered that this might be due to their treatment with phenol, and so compared the 6% gel patterns of SDS-treated polysomes, phenolated total RNA, SDS-treated 14 S-mRNP and phenolated 9S RNA from 14 S mRNP. There was no difference at all in the patterns, nor in the activity of SDS-prepared and phenol-prepared 9S RNA in cell-free protein synthesis systems nor in DNA/RNA hybridisation behaviour.

* *Dr. Williamson, as Chairman of the Symposium session, opened his remarks by regretting that Dr. Samarina [cf. Article 15] had not been allowed to attend, for which reason he made some general observations now set down.*

Why is this negative result worth reporting at all? Recently there has been a great deal of publicity given to poly-A sequences in mRNA (6-9) There is thought to be an A-rich sequence in globin mRNA (10). However, it has been suggested that poly-A sequences are extracted into the phenol phase or interphase under some conditions, and that the use of phenol/chloroform rather than phenol/cresol avoids this artifact. Our experiments demonstrate that, at least in the case of globin mRNA, phenol does not affect the size or properties of the mRNA, nor seem to affect its secondary structure, though we are less sure on this point. However, there is now a great need for a careful study, of the sort which I am sure Kirby would have performed had he been alive, on the selective removal of poly-A regions of mRNA and HnRNA by phenol and a comparison of Darnell's phenol/chloroform technique (8,9) with the more usual phenol/cresol mix.

Other questions which must be answered relate to the protein which is still associated with RNA after some phenol procedures. Is this non-specifically bound, denatured protein, or a specific protein which is very tightly bound and resists dissociation? Similarly, now that the roles of different nuclear RNAs are becoming clearer, it is necessary to reinvestigate the significance of the different fractions which are obtained by phenol extraction under different ionic conditions and temperatures (11). Finally, because of variable yields and the risk of entrapment of RNA with denatured protein, Noll and Stutz (12) have even recommended the abandonment of phenol techniques and the use of SDS alone for nucleic acid isolation. There is considerable evidence that non-ribosomal RNAs in particular can be trapped with protein in the interface selectively, at least from some tissues (13).

With reference to some of Samarina and Georgiev's experiments, in their absence I merely comment that their results with formaldehyde-fixed and unfixed mRNP preparations from mammalian nuclei indicate the presence of mRNA sequences in large heterogeneous nuclear RNA molecules, both from hybridisation data and from preliminary end-group analysis. This nuclear mRNA precursor is associated with a protein which migrates in gels as a single band, and which can be denatured and removed with phenol. This protein is different from those found associated with mRNA when it is dissociated from polysomes with EDTA (14).

The isolation of messenger RNAs for several proteins (globin, histones, immunoglobulin, silk fibroin, lens crystallin, ovalbumen) is either proven or moving ahead rapidly. This availability of biologically active RNA molecules from animal systems will permit a functional and relevant investigation of isolation techniques, not in the circumstantial way previously possible, but in more meaningful terms.

Acknowledgement

This work has been supported by grants to the Beatson Institute for Cancer Research, from the Medical Research Council and the Cancer Research Campaign.

References

1. Kirby, K.S., *Biochem J. 67* (1956) 405.
2. Kirby, K.S., *Biochem J. 96* (1965) 266
3. Parish, J.H. & Kirby, K.S., *Biochim. Biophys. Acta 129* (1966) 557.
4. Williamson, R., Morrison, M., Lanyon, G., Eason, R. & Paul, J., *Biochemistry 10* (1971) 3014.
5. Williamson, R. & Morrison, M., *Proceedings of C.N.R.S. Symposium on Molecular Pathology,* Paris, 1971, in press.
6. Edmonds, M., Vaughan, M.H. & Nakazato, H., *Proc. Nat. Acad. Sci. (Wash.) 68* (1971) 1336.
7. Lee, S.Y., Mendecki, J. & Brawerman, G., *Proc. Nat. Acad. Sci. (Wash.) 68* (1971) 1331.
8. Darnell, J.E., Wall, R. & Tushinski, R.J., *Proc. Nat. Acad. Sci. (Wash.) 68* (1971) 1321.
9. Philipson, L., Wall, R., Glickman, G. & Darnell, J.E., *Proc. Nat. Acad. Sci. (Wash.) 68* (1971) 2806.
10. Lim, L. & Canellakis, E.S., *Nature (Lond.) 227* (1970) 71.
11. e.g. Sibatini, A., deKloet, S.R., Allfrey, V.G. & Kirsky, A.E., *Proc. Nat. Acad. Sci. (Wash.) 48* (1962) 471.
12. Noll, H., & Stutz, E., *Methods in Enzymology* (Colowick, S.P. & Kaplan, N.O., eds.), *12B* (1968) 129.
13. Barlow, J. & Mathias, A.P., in *Procedures in Nucleic Acid Research* (Cantoni, G.L. & Davies, D.R., eds.), Harper & Row, London (1967), p.444
14. Lukanidin, E.M., Georgiev, G.P. & Williamson, R., *FEBS Lett. 19* (1971) 152.

15 THE HOT PHENOL EXTRACTION OF RNA, PARTICULARLY NUCLEAR dRNA

O.P. Samarina, V.L. Mantieva, A.P. Ryskov and G.P. Georgiev
Institute of Molecular Biology
Academy of Sciences of the USSR
Moscow, USSR

Extraction of ascites cells or tissue homogenates by 0.14 M NaCl-phenol mixture, pH 6, at stepwise-elevated temperatures allows one to isolate: (1) cytoplasmic RNA; (2) nucleolar rRNA, including all rRNA precursors; (3) chromosomal dRNA (synonyms: nuclear Hn-RNA; ml-RNA). The cross-contamination of nuclear dRNA and rRNA fractions does not exceed 10%, as is evident from the data on base composition and alkaline-stable dinucleotide content. It is possible to obtain different fractions of nuclear dRNA in an almost non-degraded state. Results on the structure of newly formed dRNA are presented.

In 1959-60 Sibatani *et al.* (1) and Georgiev and Mantieva (2,3) found that treatment of a cell suspension with phenol at pH 6 failed to remove about one-tenth of the total RNA, which remains in the interphase formed between the water and phenol phases. The interphase layer contained unbroken cell nuclei (so-called 'phenolic nuclei'). RNA recovered in the interphase layer was identified as chromosomal and nucleolar RNA (3). It was extracted from the 'phenolic nuclei' by means of treatment with 0.14 M NaCl-phenol mixture at 65° and shown to be a mixture of rRNA and dRNA (RNA with DNA-like base composition) (4).

These two kinds of RNA could be partially separated by various methods (5,6). The most effective is the hot phenol fractionation procedure, viz. the extraction of 'phenolic nuclei' at stepwise increasing temperatures.

It was the elaboration of the hot phenol fractionation technique in 1961 that led to the discovery of the existence of DNA-like RNA in animal cells (5). The properties of material thus isolated were described in detail (6,7). A synonym for dRNA is heterogeneous or heterodisperse RNA (Hn-RNA); this term was introduced by authors who re-described the same RNA four years later (8-10). Another name suggested for this RNA is messenger-like RNA (ml-RNA) (11,12).

Hot phenol fractionation allows one to separate fractions of nuclear dRNA which probably differ in maturation state. It also allows one to isolate undegraded nucleolar RNA (14) which may serve as a source for the further isolation of pre-rRNAs (15,16). Finally hot phenol fractionation is a very simple procedure for the preparation of undegraded crude cytoplasmic RNA (6).

METHODS

The idea of the hot phenol fractionation is to avoid preliminary isolation of the subcellular structures (i.e. nuclei etc.) during which enzymatic degradation of RNA may take place. The addition of the phenol at the first stage of procedure should inhibit this process.

We used hot phenol fractionation for the isolation of RNA fractions from rat liver and Ehrlich ascites carcinoma cells. In both cases it works well. Some authors successfully applied the technique to other tissues (17-19). There are also claims in the literature that this method does not work even in the case of rat liver (20). This question will be discussed in more detail below, but in general the inability to reproduce the technique has been due to modifications arbitrarily introduced into the procedure by the authors concerned (20). This was clearly demonstrated in a recent paper from Tsanev's laboratory (21) as the outcome of careful checking of the method.

Isolation of RNA fractions from Ehrlich ascites carcinoma cells (6,22,23)

The following reagents are required:- 0.14 M NaCl phenol, pH 6 (freshly distilled phenol saturated with water and adjusted to pH 6 with NaOH); 0.14 M NaCl-1% sodium dodecyl-sulphate (SDS); 20% SDS-chloroform (distilled).

Ehrlich ascites carcinoma cells are collected by centrifuging and washed once with 0.14 M NaCl. Then they are suspended in cold 0.14 M NaCl by intensive shaking (the ratio of the volume of packed cells to 0.14 M NaCl being 1:10 to 1:20) and an equal volume of phenol pH 6 (likewise at 4°) is added. The mixture is shaken for 15 min at 4° and then centrifuged for 10-15 min at 5000 rev/min or for 20 min at 3000 rev/min. Three layers are formed: water layer at top, white interphase, and phenol layer; also there is a sediment at the bottom of the tube. The water layer is collected carefully ('cold fraction'). The interphase (crude 'phenolic nuclei') is transferred into another vessel and mixed with phenol pH 6 and 0.14 M NaCl in equal amount. The total volume is the same as during the first treatment.

The shaking and centrifugation are repeated. The water layer is rejected or combined with the first extract. The interphase is again collected, and treated once or twice more in exactly the same way but with a reduced amount of saline-phenol mixture (half the original volume). The number of cold phenol treatments influences the purity of the next fraction (nucleolar RNA), and this requirement determines the extent to which purification is undertaken during the initial steps.

Then the phenolic nuclei are again mixed with the phenol, pH 6, and an equal volume of 0.14 M NaCl. The volume of each solvent has to be

10-20 times the volume of packed phenolic nuclei. The mixture is intensively shaken at 40° for 15 min in a water bath, then cooled under tap water and centrifuged for 10 min at 5000 rev/min at 2°. The water layer is collected (40° fraction) and the same amount of 0.14 M NaCl is added to the interphase + phenol. Then the material is transferred to the vessel for shaking, and is shaken for 15 min at 55°. After cooling and centrifuging, the water layer is collected (55° fraction) and again an equal amount of 0.14 M NaCl is added. The mixture is shaken in the cold for 5-10 min and centrifuged, and the water phase is rejected or combined with the 55° fraction.

At this step the interphase is separated from phenol carefully, suspended in 10-20 volumes of 0.14 M NaCl - 1% SDS and shaken vigorously. Then an equal volume of phenol, pH 6, is added and the mixture is shaken for 15 min at 65°. After cooling and centrifugation for 15 min at 5000 rev/min (at 2°) the water phase is collected (65° fraction). The interphase is also collected, suspended in the same volume of 0.14 M NaCl - 1% SDS, mixed with phenol, and shaken for 15 min at 85°. Then it is cooled and centrifuged, and the aqueous layer is collected (85° fraction).

Immediately after their separation, all RNA fractions are treated in the same way. After addition of 0.05 vol. of 20% SDS (except to the 65° and 85° fractions which already contain it) and an equal volume of phenol, pH 6, the mixture is shaken for 10 min in cold water. After 15 min centrifugation at 5000 rev/min at 2° the clear aqueous layer is collected, shaken with an equal volume of chloroform for 10-15 min and centrifuged for 15 min. Sometimes if the water phase is not quite clear the chloroform treatment is repeated. Thereafter the aqueous layer is carefully collected, 2.5 vol of distilled ethanol are added to precipitate RNA, and the mixture is left overnight in the cold.

The RNA fractions are collected, dissolved in a small volume of water, and precipitated by the addition of 2.5 volumes of ethanol and 0.1 volumes of 2 M sodium acetate, pH 7.0. The precipitates are dissolved in 0.5-1.0 ml of water. $MgCl_2$ and Tris-HCl, pH 8.0 are added to final concentrations of 0.001 M and 0.1 M respectively. After addition of 5-20 µg of DNase (free from any traces of RNase; Worthington, electrophoretically pure), the sample is incubated for 20 min at 25°. The reaction is stopped by addition of EDTA to 0.005 M and SDS to 0.5%. Then the sample may be layered on the top of a sucrose density gradient (50 ml of 5-20% sucrose in 0.05 M NaCl-0.005 M EDTA-0.01 M Tris pH 7.5-0.25% SDS; usually 5 ml of 50% sucrose is put underneath the gradient). It is centrifuged in the SW-25.2 rotor of a Spinco L-2 ultracentrifuge at 20° either for 6 h at 24,000 rev/min or for 14 h at 16,000 rev/min. Altogether 20-25 fractions are collected, and after UV absorbance and radioactivity have been measured on aliquots, they are combined and precipitated with ethanol, whereafter they are ready for use. With this step the degraded DNA is removed (it stays on the top of gradient), and the RNA products may be further subfractionated on the basis of their molecular weights.

Isolation of RNA fractions from tissue material

This is essentially as for the ascites cells, except that before the addition of phenol the tissue is homogenized in 0.14 M NaCl. If, for example, rat liver is to be used, it is minced by scissors, mixed with 20 vol. of cold 0.14 M NaCl and homogenized by several vertical strokes of a loosely fitting Potter-Elvehjem homogenizer. Each sample of homogenate is immediately filtered through two layers of gauze and mixed with an equal amount of phenol. The subsequent steps are exactly as with ascites cells.

Nature of the fractions obtained

In studying the RNA fractions by gradient centrifugation procedures, or by electrophoretic analyses as illustrated in Fig. 1, it is useful to pre-label the RNA *in vivo*, suitably for 30 min with ^{14}C-orotic acid or for 60 min with ^{32}P-phosphate. With the latter (100 μC per animal), the ^{32}P distribution amongst the 2'(3')-nucleotides finally isolated furnishes base composition values for newly formed RNA.

In the cold fraction most of the cellular RNA is recovered, including all the cytoplasmic ribosomal RNA and also cytoplasmic mRNA (1-3). It is not yet proved that all cytoplasmic mRNA is extracted at this stage of the procedure. It should be pointed out, however, that after short pulses or in the presence of Actinomycin at low concentration, the labelled high-molecular weight RNA of this fraction has a DNA-like base composition. In many respects it is very similar to the rapidly labelled RNA of polysomes (24, 25).

The 40° fraction contains RNA of nucleolar origin including all pre-rRNA (6, 26, 27) and some mature rRNA (Fig. 1*). These RNAs are characterized by high G + C content (Table 1). All rapidly labelled high-molecular weight pre-rRNA is definitely the nucleolar component. Some mature rRNA, notably 18 S rRNA, may represent cytoplasmic contamination. Its content becomes lower if the number of cold extractions is increased.

The 55° fraction is a mixture of rRNA and dRNA (Fig. 1 & Table 1). Its overall base composition is intermediate, the ratio G + C to A + U being near unity. The 65° and 85° fractions contain dRNA. The specific activity of the 85° fraction after short labelling is significantly higher. The 65° fraction is rich in relatively low molecular weight material of 18 S type, which is labelled rather slowly (7, 22).

Cytoplasmic RNA, nuclear rRNA and dRNA may be obtained almost quantitatively by the technique described. In the case of rat liver the yield from 1 g of tissue amounts to 0.05 mg of nuclear rRNA and 0.05 mg of nuclear dRNA. From Ehrlich carcinoma cells the yield is higher, due to the higher content of nuclear material.

* *For discussion of Fig. 1, see later in the text*

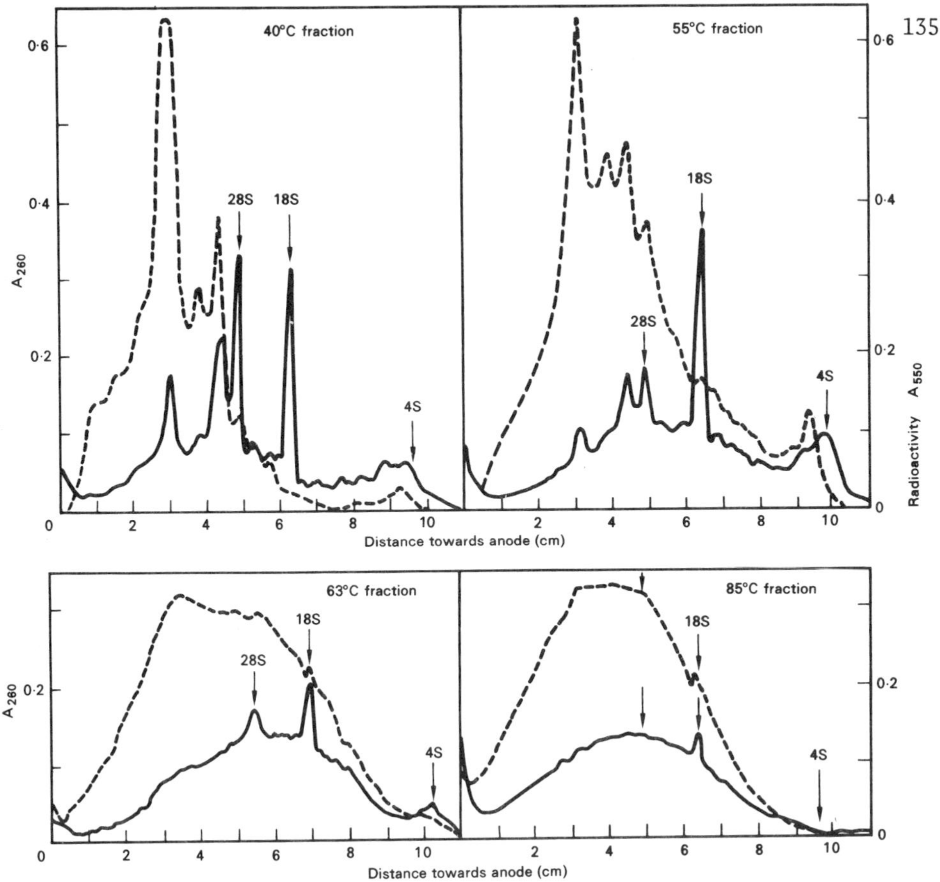

Fig. 1. Electrophoregrams of RNA fractions isolated from Ehrlich ascites carcinoma cells. *Electrophoresis was performed in agar gel according to the technique of Markov & Arion [21] (similarly to Fig. 1 of Article 6 in this book* - Ed.). *The broken line signifies radioactivity (arbitrary units based on* A_{550} *readings on 'autoradiographs').*

Variants of the method

Depending on the aims of the experiment the technique may be changed, mainly in respect of the choice of temperature intervals. We present here two variants of the method.

(1) The isolation of total nuclear dRNA without special attention to its purity.– Phenolic nuclei are treated at 40° and then extracted directly at 85° twice, in the presence of SDS.

(2) Improved separation of the rather slowly labelled and rapidly labelled dRNA fractions.– Phenolic nuclei are treated at 40° and at 55°, washed, and then treated at 63-65° without SDS (slowly labelled dRNA),

whereafter they are extracted twice with SDS present at 85° (rapidly labelled dRNA).

Many other variants may be used. It is important, however, that SDS be introduced into the extraction medium only at the moment when one does not need any further fractionation and wants only to isolate all dRNA remaining in the phenolic nuclei.

Basis of the method

Unfortunately the hot phenol fractionation procedure is still empirical. We do not know the reasons for the liberation of particular RNA fractions at particular temperatures. Probably the difference in response of nuclear and cytoplasmic RNA species to the phenol treatment depends on the structure of corresponding ribonucleoprotein complexes. It is known that all the nuclear dRNA is combined with special protein particles, informofers, probably being distributed over their surface (28). In cytoplasm the informofers and dRNA may determine the requirements of heating for deproteinization. Moreover, nucleolar rRNA is combined with some proteins which probably are absent from cytoplasm (29). However, final conclusions await special model experiments.

PURITY AND CHARACTERISTICS OF THE RNA FRACTIONS OBTAINED

Base composition

The base composition is one of the main checks on the dRNA or rRNA purity. Fortunately in most animals total DNA and rRNA have very different base compositions. The ratio G + C to A + U is 0.75 for DNA and 1.6 for rRNA. Therefore it is possible to calculate roughly the cross-contamination of RNA fractions on the basis of base composition.

In Table 1 the base composition of RNA fractions obtained by the hot phenol method is presented. The 40° fraction contains RNA with the ratio G + C to A + U = 1.6-1.7. The same result was obtained with either of the two techniques, viz. direct measurement of UV absorbance in separated nucleotides, and measurement of ^{32}P in the spots containing 2'- and 3'-nucleoside monophosphates. Thus both the total and the newly formed RNA of the fraction belong to the rRNA type by the base composition criterion. On the other hand, the fractions isolated at 65° and 85° have a base composition of AU-type, viz. G + C to A + U = 0.75-0.9. Again the two techniques give the same result (Table 1).

On calculating the purity of the fractions on the basis of the above mentioned coefficients for pure rRNA and DNA, one finds that dRNA contains less than 10% of rRNA, whereas rRNA contains less than 5% of dRNA.

Another typical difference is the composition of the RNA species studied is the presence in rRNA of alkali-stable dinucleotides. This is

Table 1. Base composition of dRNA fractions obtained with the aid of hot phenol fractionation (22)

Material	Fraction	S_{20}	Base composition of total RNA					Base composition of newly formed RNA				
			A	U	G	C	$\frac{G+C}{A+U}$	A	U	G	C	$\frac{G+C}{A+U}$
Rat	0°	-	17.0	20.9	33.9	28.2	1.64	-	-	-	-	-
liver	40°	-	17.5	23.8	31.0	27.7	1.42	17.8	19.7	34.3	28.3	1.67
	55°	-	23.5	26.9	27.1	22.5	0.99	24.2	26.1	26.7	22.9	0.99
	63°	-	25.0	25.7	25.6	23.6	0.97	28.7	23.0	24.6	23.7	0.97
	85°	-	24.5	30.5	23.5	21.5	0.82	28.6	25.7	22.6	23.2	0.84
Ehrlich	0°	-	19.3	19.6	31.9	29.1	1.57	-	-	-	-	-
carci-	40°	-	17.9	21.8	30.7	29.6	1.52	14.5	21.7	30.6	33.2	1.77
noma cells	55°	-	21.3	23.7	28.0	26.7	1.20	17.6	25.2	28.2	29.0	1.33
	63°	-	29.1	24.6	23.3	23.1	0.86	27.7	24.7	21.1	23.6	0.81
	85°	-	27.0	28.7	24.3	20.1	0.80	28.4	28.1	21.5	21.9	0.77
Ehrlich	40°	18S	-	-	-	-	-	14.8	23.4	27.8	34.0	1.62
carci-		23S	-	-	-	-	-	13.9	22.2	30.6	33.2	1.77
noma		28S	-	-	-	-	-	15.6	22.2	29.3	32.2	1.65
cells		32S	-	-	-	-	-	14.4	22.5	31.8	31.4	1.71
		45S	-	-	-	-	-	14.5	21.1	32.0	32.6	1.81
		>45S	-	-	-	-	-	15.3	21.3	31.4	32.0	1.73
	63°	<18S	-	-	-	-	-	34.0	23.9	19.8	22.2	0.73
		18S	28.6	24.4	23.3	23.7	0.89	31.3	23.8	19.3	25.6	0.81
		23S	-	-	-	-	-	27.7	25.7	21.8	24.8	0.87
		28S	-	-	-	-	-	28.0	26.3	22.1	23.5	0.83
		32S	-	-	-	-	-	27.4	27.1	22.1	23.4	0.83
		45S	-	-	-	-	-	26.3	28.6	21.9	23.3	0.80
		>45S	-	-	-	-	-	26.1	27.9	21.6	24.3	0.85
	85°	≤18S	-	-	-	-	-	26.1	26.4	22.3	25.2	0.90
		18-45S	-	-	-	-	-	26.4	27.9	21.5	24.1	0.84
		>45S	-	-	-	-	-	26.4	29.1	21.3	23.2	0.80

the result of methylation of several ribose residues in the rRNA, giving 2'-O-methylribose; dinucleotides containing it cannot be hydrolyzed by alkali. Only those sequences in the rRNA precursor are methylated which are converted subsequently into 28 S and 18 S rRNA. The sequences being degraded do not contain 2'-O-methylribose. Therefore mature rRNA contains a higher proportion of stable dinucleotides than pre-rRNA (30, 31). On the other hand it is known that mRNA does not contain alkali-stable dinucleotides. For this reason the content of the latter in dRNA may be used as a marker of contamination with rRNA. They comprise about 0.05% of total ^{32}P labelling. Thus the contamination of dRNA with rRNA does not exceed 5-10%.

Recently it was found that nuclear dRNA contains long poly-A sequences (100-200 nucleotides) which comprise about 0.5% of ^{32}P labelling in dRNA (32-34). To isolate these sequences, RNA is degraded by pyrimidine- and guanine-specific RNases in a medium of fairly high ionic strength (0.1-0.3 M salt) and then chromatographed on Sephadex G-75 (35). A very sharp peak of acid-insoluble material was found with the dRNA digest. It was localized between the positions of rRNA and tRNA. About 90% of bases in the peak consisted of A. Then RNA isolated at 40° was analyzed for the presence of poly-A. Not even trace amounts of long poly-A sequences were found. Thus according to this test the 40° fraction does not contain dRNA at all.

Finally, in the course of 3'-end analysis we found a new example of a marked difference between 40°, 65° and 85° fractions. The 3'-ends were oxidized with periodate and then reduced with sodium borohydride. After alkaline hydrolysis the labelled 3'-nucleotides were purified, separated and counted. The 40° fraction (30-50 S material) was found to contain C and U. No A was found. On the other hand, dRNA fractions contain 75% A, 20% U and only 5-10% of G + C (Table 2). Thus again the cross-contamination of rRNA and dRNA does not exceed 5-10%.

Thus all the above-mentioned tests confirm the good separation of nuclear rapidly labelled RNA species by the hot phenol fractionation technique. The reason why some authors have not achieved such separation of dRNA is that they did not follow the description but introduced some modifications (20).

Quality of the RNA fractions in relation to possible degradation

It is very important to know whether the technique does or does not allow one to isolate RNA in a non-degraded state. From experiments in which total RNA was isolated directly from the cell, it is known that newly formed rRNA and dRNA have a very high molecular weight. The main components of pre-rRNA have sdeimentation coefficients of 45 S and 32 S (15, 16), and newly formed dRNA sediments in the region between 20-30 and 100 S (8-11). The molecular weights of RNA fractions may be determined with the aid either of sedimentation or of electrophoresis in polyacrylamide or agar gel.

Table 2. Base composition of 3'-ends of nuclear dRNA isolated from rat liver

Nature of dRNA fraction	END NUCLEOSIDE, *as % of total 3'-ends*			
	A	U	G	C
Total	70	20	4 +	6
Heavy	71	13	16	
Intermediate	77	14	9	
Light	68	19	13	

In the 40° fraction obtained by our method, all rapidly labelled RNA is concentrated in 45 S peaks and also, after long-term labelling, in 32 S peaks (Fig. 1). Thus no degradation of rRNA takes place during the 40° treatment. In the UV absorbance curve one can also find 28 S and small 18 S peaks, but they do not contain label after pulse labelling. Thus the 40 fraction contains undegraded rRNA precursors (26).

Electrophoregrams of pulse-labelled dRNA fractions are also presented in Fig. 1. In the 65° fraction much UV-absorbing material is localized in a relatively low molecular weight zone – about 10-25 S (peak at 18 S) – although some material is also revealed in a heavier zone of 30-70 S. However, the radioactive material after pulse labelling has a much higher molecular weight – the peak being in the 30-40 S zone, and a significant proportion of the material being even heavier (up to 70-100 S).

The 85° fraction is metabolically more homogeneous: both the UV-absorbing and the radioactive material show a broad peak in the 30-40 S region, and a heterogeneous distribution from 20 to 100 S (14, 23).

Thus newly formed nuclear dRNA isolated by the hot phenol fractionation technique has a very high molecular weight ($2\text{-}10 \times 10^6$), of the same order as that of dRNA isolated directly from the cells without preliminary fractionation. Of course, the possibility cannot be excluded that there may have been some degradation, especially in the case of the 85° fraction; but such degradation is insignificant.

Additional evidence in favour of the absence of any significant degradation comes from the end analysis of dRNA (36, 37). It should be pointed out that neither phosphorylated 5'-ends nor nucleosides from the 3'-ends can appear in the course of non-specific degradation. It was found at the outset that heavy dRNA (35 S or greater) contains triphosphorylated or monophosphorylated 5'-ends, whereas the light (10-20 S) and intermediate (20-30 S) fractions contain only monophosphorylated 5'-ends (Table 3). The absence of pppXp groups in alkaline hydrolysates of dRNA with sedimentation coefficients less than 35 S shows that the difference in molecular weight is true and is not a result of degradation of high-molecular weight RNA during the isolation. It is possible to make rough calculations of molecular weights on the basis of the content of the terminal phosphate groups. Such calculation shows that the molecular weight of dRNA of 35 S

Table 3. Nature of 5'-ends in different rat liver dRNA fractions

The mol. wt. figures are calculated on the basis of assumption that the specific activities of γ, β- and α-phosphates in RNA-nucleotides are the same. This should lead to some underestimation of average mol. wt. of heavy dRNA, containing tri-phosphorylated 5'-ends, as δ and β-phosphates are in fact labelled more rapidly.

No. of experiment	S_{20} of fraction	dRNA mol. wt. $\times 10^{-6}$	Xp	RADIOACTIVITY pppXp + ppXp		pXp		Calculated average molecular weight
			counts/min $\times 10^{-6}$	counts / min	% of Xp	counts /min	% of Xp	
1	>30	>2	1.44	590	0.041	187	0.013	2×10^6
	10-30	0.2-2.0	0.6	25	0.004	600	0.10	0.64×10^6
2	>35	>2.5	1.65	360	0.022	155	0.009	3.3×10^6
	20-30	0.7-2.0	0.7	< 10	<0.002	200	0.030	2.1×10^6
	10-18	0.2-0.7	0.57	< 10	<0.002	250	0.044	1.4×10^6

or greater is about 3.5×10^6 (Table 3), that of 20-30 S dRNA is 1.8×10^6, and that of 10-20 S is 1.0×10^6. These figures (except for the last, which is somewhat higher than might be expected) are in very good agreement with the molecular weight estimations made on the basis of sedimentation coefficients.

The analysis for 3'-ends did not allow the molecular weight to be estimated, but it did demonstrate that the content of 3'-ends in heavy dRNA is much lower than in light dRNA (Table 3).

Thus the end-analysis gives evidence that the rapid sedimentation of heavy material really reflects a high molecular weight. Finally a clear difference between heavy and light dRNA species was observed on investigating poly-A. Its content in heavy dRNA is much lower than in light dRNA (35). There is also a difference in the base composition of hybridizable base sequences. In the light fraction they are very A-rich (40-45% A), whilst in the heavy dRNA the hybridizable sequences are enriched in both A and G (38). Thus again the light fraction is not a product of dRNA degradation during the isolation, and the heavy fraction is not formed from the light one as a result of aggregation.

One can conclude that hot phenol fractionation allows the isolation of almost undegraded nucleolar rRNA and chromosomal dRNA in a quite pure state.

Nature of nuclear RNA fraction, and terminology

It is well known that 45 S and 32 S rRNA species are the precursors of ribosomal RNA (15, 16). They may be denoted as pre-RNA_{45} and pre-$rRNA_{32}$, in contrast to mature ribosomal RNA species - $rRNA_{28}$, $rRNA_{18}$ and $rRNA_{5}$. The sequences which are degraded may be denoted as pseudo-rRNA (ps-rRNA).

The nuclear dRNA is reckoned to be a precursor of mRNA. Although this postulate is not accepted by all authors, many observations support the idea.–
(1) The data on competitive hybridization: giant nuclear dRNA completely inhibits the hybridization of polysomal mRNA with DNA, whilst polysomal RNA only partly inhibits the hybridization of nuclear giant dRNA (39-42).
(2) mRNA effectively inhibits the hybridization of the 3'-ends of giant dRNA with DNA; this also indicates that the mRNA sequence is localized near the 3'-end of giant precursor molecules (36-43).
(3) Some specific sequences such as poly-A are present in both the nuclear dRNA and mRNA.
(4) Some virus-specific sequences may be found also in cytoplasmic mRNA and in nuclear dRNA (44, 45).
(5) Nuclear dRNA and cytoplasmic mRNA have similar short polypurinic sequences at the 3'-end (46, 47).

Thus nuclear dRNA may be regarded as a precursor of mRNA and be denoted pre-mRNA.

If one wishes to indicate its size, a subscript may be used to signify sedimentation coefficients: pre-$mRNA_{30-70}$, pre-$mRNA_{18}$, etc. Again the sequences degraded inside the nucleus are referred to as pseudo-mRNA (ps-mRNA). The same type of terminology can be used for all RNA species: for example, tRNA, pre-tRNA, ps-tRNA.

References

1. Sibatani, A., Yamana, K., Kimura, K. & Okagaki, H., *Biochim. Biophys. Acta 33* (1959) 590.
2. Georgiev, G.P. & Mantieva, V.L., *Biokhimiya 25* (1960) 143.
3. Georgiev, G.P., *Biokhimiya 24* (1959) 472.
4. Georgiev, G.P., *Biokhimiya 26* (1961) 1095.
5. Georgiev, G.P. & Mantieva, V.L., *Vopr. Med. Khimii 8* (1962) 93.
6. Georgiev, G.P. & Mantieva, V.L., *Biochim. Biophys. Acta 61* (1962) 153; *Biokhimiya 27* (1962) 949.
7. Samarina, O.P., *Biochim. Biophys. Acta 91* (1964) 688.
8. Warner, J.R., Soeiro, R., Birnboim, C., Girard, M. & Darnell, J.E., *J. Mol. Biol. 19* (1966) 349.

9. Penman, S., Vesco, R. & Penman, M., *J. Mol. Biol. 34* (1968) 49.

10. Attardi, G., Parnas, H., Huang, M.I.H. & Attardi, B., *J. Mol. Biol. 20* (1966) 145.

11. Scherrer, K. & Marcaud, L., *Bull. Soc. Chim. Biol. 47* (1967) 1967.

12. Scherrer, K. & Marcaud, L., *J. Cell Physiol. 72, Suppl. I* (1968) 181.

13. Ryskov, A.P. & Georgiev, G.P., *FEBS Lett. 8* (1970) 186.

14. Georgiev, G.P. & Lerman, M.I., *Biochim. Biophys. Acta 91* (1964) 678.

15. Perry, R.P., *Proc. Nat. Acad. Sci. (Wash.) 48* (1962) 2179.

16. Scherrer, K., Latham, H. & Darnell, J.E., *Proc. Nat. Acad. Sci. (Wash.) 49* (1963) 240.

17. Zimmerman, E. & Turba, F., *Biochem. Z. 339* (1964) 469.

18. Mach, B. & Vassalli, P., *Science (Wash.) 150* (1965) 622.

19. Samis, H.F., Wulff, V.J. & Falzone, J.A., *Biochim. Biophys. Acta 93* (1964) 223.

20. Hadjiolov, A., *Progr. Nucleic Acid Res. & Mol. Biol. 7* (1968) 196.

21. Markov, G. & Arion, V. Ya., *Eur. J. Biochem.*, in press.

22. Arion, V. Ya., Mantieva, V.L. & Georgiev, G.P., *Molekul. Biol. (USSR), 1* (1967) 689.

23. Mantieva, V.L., Avakjan, E.R. & Georgiev, G.P., *Molekul. Biol. (USSR) 5* (1971) 321.

24. Samarina, O.P., *Dokl. Akad. Nauk SSSR 156* (1964) 1217.

25. Tsanev, R., *Biochemistry of Ribosomes and Messenger RNA* (R. Lindigkeit, P. Langen & J. Richter, eds.) Akad. Verlag, Berlin (1968), p. 293.

26. Lerman, M.I., Vladimirzeva, E.A., Terskich, V.V. & Georgiev, G.P., *Biokhimiya 30* (1965) 375.

27. Georgiev, G.P., Samarina, O.P., Lerman, M.I. & Smirnov, M.N., *Nature (Lond.) 200* (1963) 1291.

28. Georgiev, G.P. & Samarina, O.P., *Adv. Cell Biol. 2* (1971) 47.

29. Maden, B.E.H., *Nature (Lond.) 219* (1968) 685.

30. Lane, B.G. & Tamaoki, T., *J. Mol. Biol. 27* (1967) 335.

31. Chei, Y.C. & Busch, H., *J. Biol. Chem. 245* (1970) 1954.

32. Edmonds, M., Vaughan, M.H. & Nakazoto, H., *Proc. Nat. Acad. Sci. (Wash.) 68* (1971) 1336.

33. Lee, S.Y., Mendecki, J. & Brawerman, G., *Proc. Nat. Acad. Sci. (Wash.) 68* (1971) 1331.

34. Darnell, J.E., Wall, R. & Tushinski, J.J., *Proc. Nat. Acad. Sci. (Wash.) 68* (1971) 1321.

35. Ryskov, A.P., Farashyan, V.R. & Georgiev, G.P., *FEBS Lett. 20* (1972) 355.

36. Coutelle, Ch., Ryskov, A.P. & Georgiev, G.P., *FEBS Lett. 12* (1970) 21; *Molekul. Biol. (USSR) 5* (1971) 334.

37. Ryskov, A.P., Farashyan, V.R. & Georgiev, G.P., *Molekul. Biol. (USSR)* in press (1972).

38. Besson, J., Farashyan, V.R. & Ryskov, A.P., *Cell Differ.* in press (1972).

39. Georgiev, G.P., in *The Cell Nucleus: Metabolism and Radiosensitivity* (M.G. Ord, L.A. Stocken, H.M. Louwen & I. Betel, eds.), Taylor & Francis, London (1966), p. 79.

40. Arion, V. Ya. & Georgiev, G.P., *Dokl. Akad. Nauk SSSR 172* (1967) 716.

41. Georgiev, G.P., Ryskov, A.P., Coutelle, Ch., Mantieva, V.L. & Avakjan, E.R., *Biochim. Biophys. Acta 259* (1972) 259.

42. Shearer, R.W. & McCarthy, B.J., *Biochemistry (Wash.) 6* (1967) 283.

43. Scherrer, K. *et al., Cold Spring Harb. Symp. Quant. Biol. 35* (1970) 539.

44. Lindberg, U. & Darnell, J.E., *Proc. Nat. Acad. Sci. (Wash.) 65* (1970) 1089.

45. Acheson, N.H., Buetti, E., Scherrer, K. & Weil, R., *Proc. Nat. Acad. Sci. (Wash.) 68* (1971) 2231.

46. Barr, H. & Lingrel, J.B., *Nature New Biol. 223* (1971) 41.

47. Ryskov, A.P., Farashyan, V.R. & Georgiev, G.P., *FEBS Lett. 20* (1972) 355.

16 PURIFICATION OF PROTEINS IN PHENOL-CONTAINING SOLVENT SYSTEMS

A. Pusztai
The Rowett Research Institute
Bucksburn
Aberdeen AB2 9SB, U.K.

Progress in the investigation of the nature and properties of the water-insoluble protein constituents of cell membranes, cell walls and other structures is dependent on finding suitable solvents for their dissolution. When these are found, protein fractionation techniques analogous to those used with aqueous systems have to be devised. There are a number of chemical agents known to bring about the solubilization of protein constituents. These include aqueous solutions of urea or guanidinium salts, in high concentration, or detergents (e.g. sodium dodecyl sulphate, sodium deoxycholate, Triton X-100, etc.), and certain organic solvents (e.g. pyridine, chloroform-methanol mixtures, chloro-ethanol, etc.). With most of these solvents, however, the more common protein fractionation methods, such as salting-out, isoelectric precipitation, electrophoresis, molecular sieve or ion exchange chromatography, lead to difficulties. In our studies which are directed toward the ultimate goal of finding an alternative to handling and purifying proteins in aqueous media, we have made use of the strong interaction between proteins and phenol.

The affinity between proteins and phenol has been recognized and used extensively by a succession of chemists ever since Runge discovered phenol in 1834, and is also known to be highly specific. Thus, the well-known and widely used deproteinization methods for nucleic acids and polysaccharides are all founded on it. It has been pointed out (1) that because the extent of interaction between proteins and polyacids is determined mainly by their pH-dependent ionization state, and because of their absolute preference for phenol, proteins can be selectively extracted from natural sources at pH values where the interaction is at a minimum. Polyacids and polysaccharides present in the same tissues are, on the other hand, left behind undissolved.

Further purification on a preparative scale of these proteins is now also possible in phenol-containing solvent mixtures by a variety of protein fractionation methods. For example, it has now been established that a pH_{app}-dependent* H^+ ionization of potentially dissociable groups of proteins can occur in buffered phenol-ethanediol-water (3:2:3; w/v/v)

**'pH_{app}' is defined as the actual pH value measured, which in fact differs from the true pH value of the medium by an unknown contribution from the liquid-junction potential. Neither the sign nor the magnitude of the latter is known.*

solvent systems. This ionization proceeds to such an extent that electrophoretic separation of proteins, based mainly on charge differences, becomes feasible in practice. Protein mobility has been shown to change with changing pH_{app} values of the medium, and an apparent isoelectric point for the individual proteins can be estimated (Fig. 1). Founded on these observations, a preparative high-voltage electrophoresis method in free-flowing buffer films has been worked out (2). Furthermore, the solubility behaviour of proteins in this solvent system is, similarly to that found in purely aqueous media, dependent on the pH_{app} value of the medium (Table 1).

Table 1. Solubility of *Phaseolus vulgaris* protease in phenol-formamide-aqueous ammonium formate (1:1:1; w/v/v) mixtures at various pH_{app} values

pH_{app}	Soluble portion (in %)	Activity in soluble portion (in % of total)
2.82	36.0	32.0
3.22	37.8	36.2
3.71	25.3	18.1
4.27	14.1	7.6
6.02	11.7	8.5
7.71	43.6	93.5

A number of proteins have been shown to have a solubility minimum at their apparent isoelectric point. Fractionation based on the selective precipitation or the dissolution at discrete pH_{app} values of some proteins can be carried out in this way. In addition, fractionation of proteins dependent on their size is also possible in a phenol-containing solvent (3,4). It has been established from our studies of molecular sieve chromatography on a Bio Gel P-100 column, in phenol-acetic acid-water (1:1:1; w/v/v), of a number of well-characterized proteins, polypeptides and other compounds of known molecular weight that the following empirical relationship exists between the elution volume and the log of the molecular weight (Fig. 2):

$$\log_{10} \text{mol. wt.} = \{5.13 - (0.53 \pm 0.013) \times (V_e/V_o)\} \pm 0.059$$

This method, besides having obvious analytical importance, is also easily adaptable to preparative applications. Several proteins such as ribonuclease (5), serum albumin and α-chymotrypsinogen (6) and rabbit muscle aldolase (A. Pusztai, unpublished observation, 1967) have been recovered from phenol-containing solvents and shown to be very similar to, if not identical with, the native proteins on the basis of measurements of optical rotation, viscosity, sedimentation, ultraviolet spectroscopy, immunochemical behaviour and enzyme activity. Thus protein fractionation methods in phenol-containing solvents, such as the ones described above, are finding applications not only in the purification of insoluble and structure-bound proteins but also in the isolation of enzymes and other biologically active proteins generally.

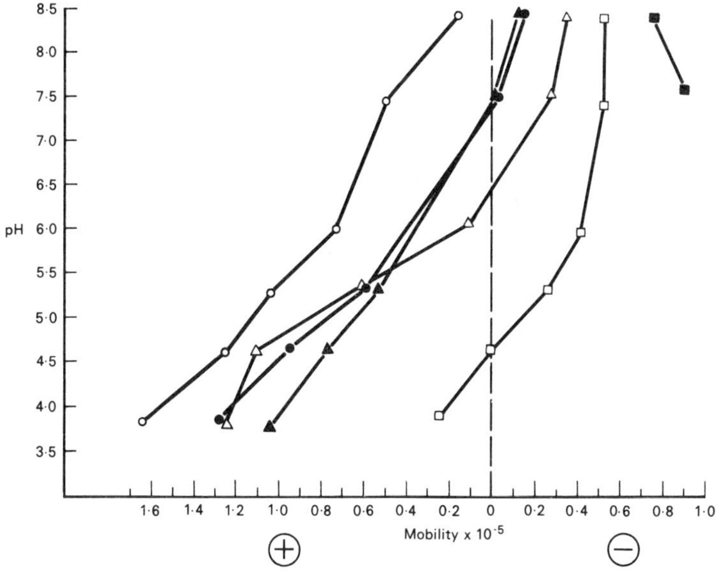

Fig. 1. The dependence of the electrophoretic mobility of proteins in buffered phenol-ethanediol-water (3:2:3; w/v/v) on the pH_{app} values of the solvent. The following proteins were investigated: cytochrome *c* (○); con-albumin (●); bovine serum albumin (△); α-chymotrypsinogen (▲); fetuin (□) and pepsin (■).

Fig. 2 *(below)*. The ratio of elution volume to void volume of substances *vs*. the log of their molecular weight value as obtained from the results of chromatography on a Bio Gel P-100 column in phenol-acetic acid-water (1:1:1; w/v/v).

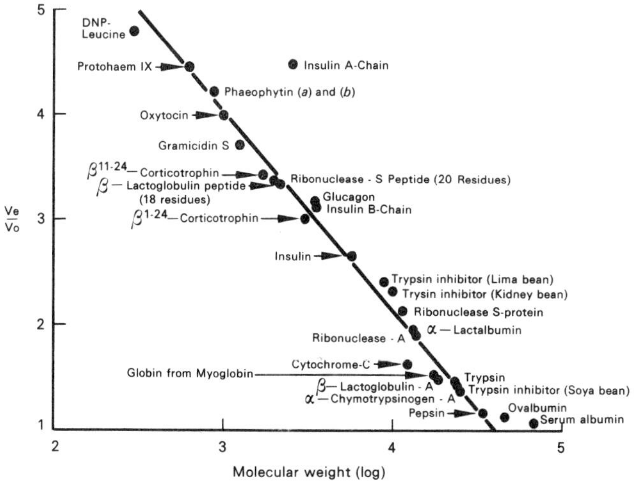

Figs. 1 & 2 have been published elsewhere [refs. 2 & 4 respectively]

ILLUSTRATIVE RUNS

The hydroxyproline-containing glycoproteins of the cell walls of *Vicia faba* represent an example of such structure-bound proteins. Because of their insolubility in ordinary aqueous solvents we attempted their purification in phenol-containing solvent mixtures. A flow-chart (Fig. 3) summarizes the main stages of our purification. Thus, the material soluble in trichloroacetic acid was resolved by phenol-aqueous (pH 8) partitioning into (a) predominantly polysaccharide-type materials with small amounts of firmly bound proteins containing hydroxyproline, and (b) into glycoproteins and glycolipids containing hydroxyproline soluble in the phenol-rich phase. These last were separated into at least four cationic, one anionic and one stationary components (Fig. 4) on free-flow electrophoresis in phenol-acetic acid-water (1:1:1; w/v/v). The stationary component contained a variety of glycolipids containing bound hydroxyproline. The four cationic components were resolved further by molecular sieve chromatography on a Bio Gel P-100 column (Fig. 5) operated in phenol-acetic acid-water (1:1:1; w/v/v). These results clearly indicated that the hypothetical cell-wall protein - extensin (7) - consists of a number of glycoproteins, polysaccharide-protein complexes and glycolipids containing hydroxyproline each with a characteristic and different composition, molecular size and solubility.

Intracellular proteolytic enzymes from plants are difficult to obtain in a pure state, especially when the purification is attempted by conventional methods and in purely aqueous media. We have, however, made some progress in the purification of one of the major intracellular proteolytic enzymes from *Phaseolus vulgaris* by a combination of fractionation methods in aqueous and in phenol-containing solvent mixtures. A scheme of the initial stages of purification is given (Fig. 6) together with the extent of the purification obtained (Table 2). Thus, after the removal of low molecular weight compounds and of basic proteins, such as ribonuclease, by extraction with and dialysis against dilute aqueous acetic acid solutions, the proteins are brought into solution by extraction with phenol-acetic acid-water (1:1:1; w/v/v). These proteins, after dialysis against aqueous acetic acid solutions and recovery by freeze-drying, are dissolved in water (at pH 8-9) and fractionated by the addition of solid ammonium sulphate. The most active fractions are precipitated at or above 40 % saturation.

The active material can be solubilized by extraction with phenol-ethanediol-water (3:2:3; w/v/v; pH_{app} 7.8) of the fractions precipitated with ammonium sulphate. An overall purification of about 130-fold is achieved at this stage. We have also made some progress in the further purification of the protease. According to the results of a number of preliminary experiments, the proteolytic enzyme can be separated successfully from the bulk of inert proteins by high-voltage electrophoresis in phenol-ethanediol-water (3:2:3; w/v/v; pH_{app} 7.8) of partially purified protease preparations (Fig. 7). The protease can also be chromatographed on our Bio Gel P-100 column in phenol-acetic acid-water (1:1:1; w/v/v).

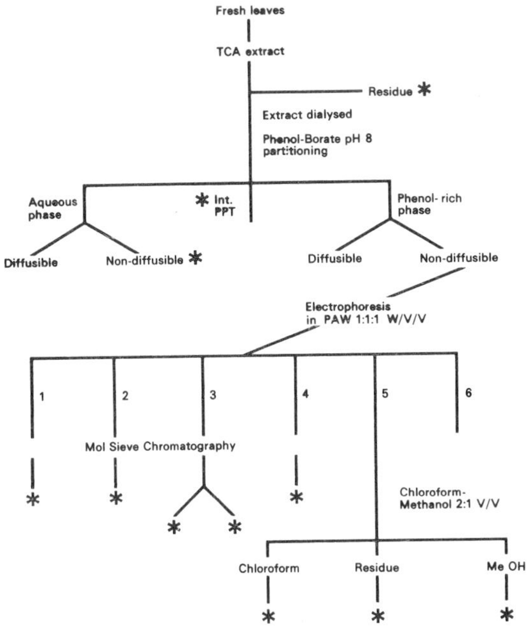

Fig. 3. Scheme of fractionation for the hydroxyproline-containing constituents of the leaves of *Vicia faba*.

Fig. 4 *(below)*. Free-flow electrophoretic separation in phenol-acetic acid-water (1:1:1; w/v/v) of materials containing hydroxyproline soluble in the phenol-rich phase on partitioning between phenol-borate (pH 8).

Figs. 3-5 have been published in ref. [3]

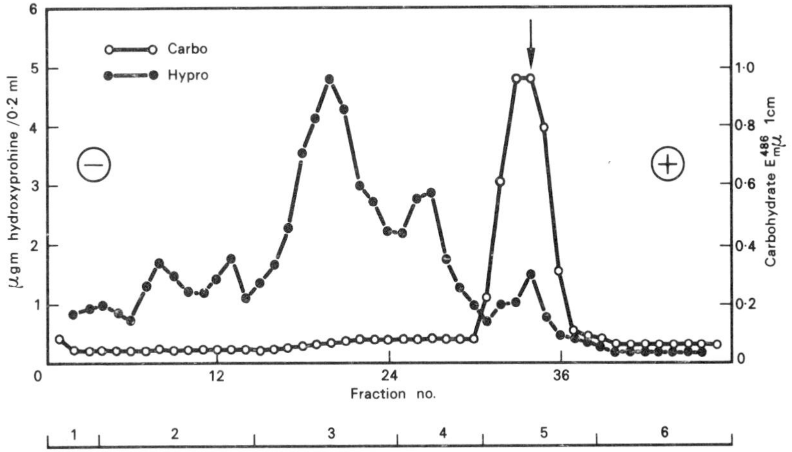

Table 2. Purification of intracellular proteolytic enzyme from the seeds of kidney bean

One activity unit = ΔA (at 280 nm)/h/g dry matter/ml reaction mixture
AS denotes ammonium sulphate.

Procedure		Amount (g)	Purity (units)	Total units	Yield (%)	Purification
Seed meal		100	14	1400	100	(1)
After extraction with 0.1 acetic acid		73.2	19	1390	99.3	1.4
Material soluble in phenol:acetic acid:water		14.3	63	903	64.5	4.5
Insoluble in aqueous buffers at pH 8-9		4.93	9.3	46	3.3	0.7
Soluble in aqueous buffers at pH 8.3	Pptd. at 0-30% AS	7.0	27.8	195	13.9	2
	Pptd. at 60-100% AS	0.78	368	287	20.5	26.3
	Pptd. at 60-100% AS	0.34	450	154	11.0	32.1
30-60% AS ppt. (0.78 g)	Pptd. at 0-40% AS	0.66	291	192	13.7	21.0
	Pptd. at 40-60% AS	0.11	871	96	6.9	62.0
60-100% AS ppt. (0.34 g)	Pptd. at 50-80% AS	0.16	678	109	7.8	48.0
	Pptd. at 80-100% AS	0.17	262	45	3.2	18.7
40-60% AS ppt. (0.11 g)	Soluble in P:ED:W; pH_{app} = 8	0.048	1786	86	6.2	128
	Insoluble portion	0.062	80	6	0.4	5.7

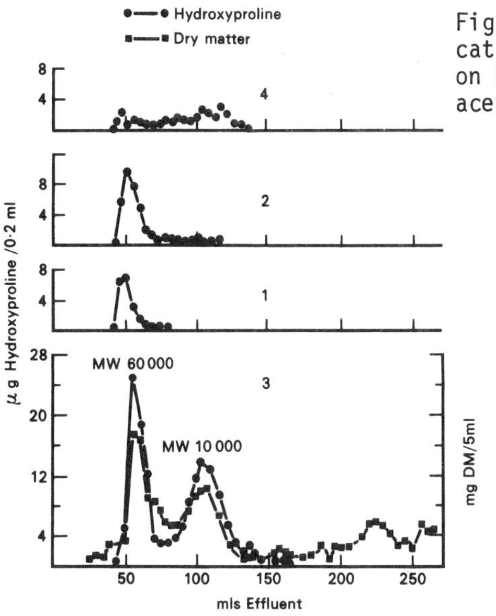

Fig. 5. Chromatography of the cationic fractions no. 1 to 4 on Bio Gel P-100 in phenol-acetic acid-water (1:1:1; w/v/v).

[*For* Fig. 6, *see overleaf*]

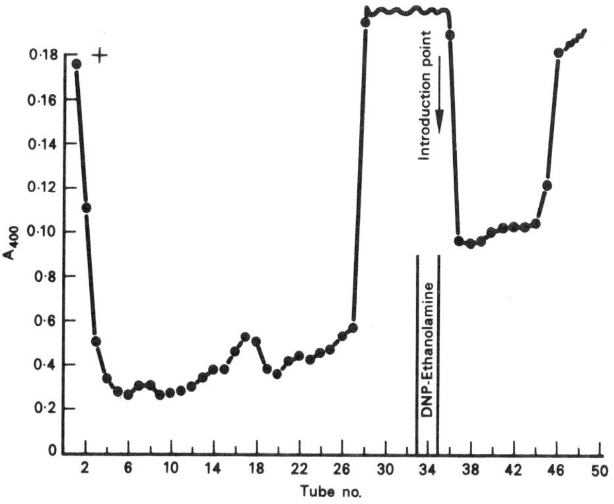

Fig. 7. Separation of kidney bean protease from inert proteins by free-flow high-voltage electrophoresis in phenol-ethanediol-water (3:2:3, w/v/v; pH_{app} 7.8).

Fig. 6. Scheme of purification of proteolytic enzyme from the seeds of kidney bean. *AS* denotes ammonium sulphate

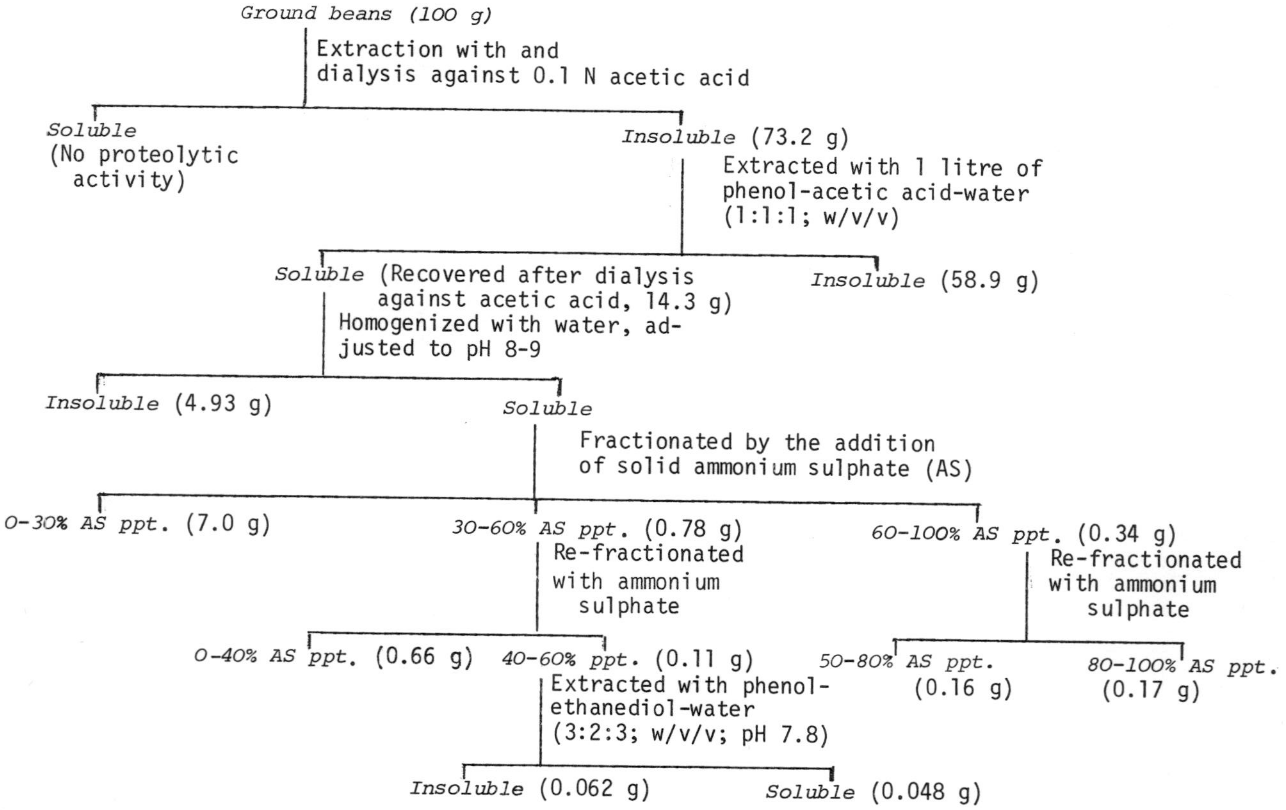

Its elution volume gave a preliminary estimate of between 20,000 and 30,000 daltons for its molecular weight. Further work based on these preliminary results is in progress to achieve a state of molecular homogeneity for this proteolytic enzyme.

References

1. Pusztai, A. *Biochem. J. 99* (1966) 93.
2. Pusztai, A. and Watt, W.B. *Biochim. Biophys. Acta 251* (1971).
3. Pusztai, A. and Watt, W.B. *European J. Biochem 10* (1969) 523.
4. Pusztai, A. and Watt, W.B. *Biochim. Biophys. Acta 214* (1970) 463.
5. Kickhöfen, B. and Bürger, M. *Biochim. Biophys. Acta 65* (1962) 190.
6. Pusztai, A. *Biochem. J. 101* (1966) 295.
7. Lamport, D.T.A., *Adv. Bot. Res. 2* (1965) 151.

17 SEPARATIONS WITH TWO LIQUID PHASES

Göte Johansson
Department of Biochemistry
University of Umeå
Sweden

The biphasic systems obtained when aqueous solutions of two suitable polymers are mixed have been used for separation of cells, viruses, cell organelles and macromolecules of biological origin [1,2,3]. The partition of substances in these biphasic systems can be varied by adding different kinds of salts or polyelectrolytes [4] or by preparing systems in which one of the polymers bears a small proportion of ionizable groups, e.g. trimethylaminopoly(ethylene glycol) [5]. It is often possible to predict qualitatively the effect of such additives on the partition in these systems of a material of known charge. The efficiency of separation can be enhanced by multistep procedures, e.g. counter-current distribution (CCD). In these biphasic systems, particles can be partitioned between the bulk phases or between the interface and one of the phases. Partition between the interface and one of the phases is frequently used when particles are to be separated. Use is made of a special CCD-apparatus, allowing a large interface and rapid settling time as well as automatic shaking and phase transfer. Particles such as chloroplasts [6,7], mitochondria [8], red blood cells [9] and Chlorella pyrenoidosa *[10] have been shown to be heterogenous when they are subjected to CCD in this type of biphasic system. Sub-populations of these particles have been isolated and shown to differ in morphological structure, age or chemical composition.*

The aqueous polymeric biphasic systems which are most used are those which are composed of the two polymers dextran and polyethylene glycol (PEG). The composition of the phases as well as the partition of added particles depend on the molecular weight of the polymers. Up to now the majority of the experiments has been done with dextran having M_w = 500,000 and with PEG having M_n = 4,000 or 6,000* (Carbowax 4000 or Carbowax 6000 obtained from Union Carbide). Phase diagrams for these and many other polymeric systems have been published (1).

* *M_w denotes weight average molecular weight, and M_n the number average molecular weight.*

$$M_w = \frac{\sum_i N_i M_i^2}{\sum_i N_i M_i} \quad \text{and} \quad M_n = \frac{\sum_i N_i M_i}{\sum_i N_i}$$

where N_i is the number and M_i is the molecular weight of the species 'i', and the summation extends over all species.

The dextran-PEG-water biphasic systems fulfil many of the conditions that are demanded of systems which should be used for CCD of biological material. –

1. The system provides an excellent environment for particles of biological origin depending on its high water content (80-95%).
2. The relatively fast settling time for these systems makes them suitable for use in CCD processes.
3. In one and the same system the partition ratio varies from one type of particle to another. Therefore the systems have in many cases a high separation capacity.
4. When a mixture of unknown composition is to be resolved in its components by CCD, it is preferable that the mean partition ratio have a value around one. This may be attained by varying the volume ratio of the two phases or by modifying the phase system. The latter can be done by addition of different kinds of salts. This technique of adjusting the partition ratio has been successfully used in many cases, but it is usually time-consuming. The volume ratio on the other hand can be varied only within narrow limits, determined by the geometry of the CCD apparatus.

By replacing a part of PEG in the system with PEG carrying ionisable groups the partition ratio for many biological particles can be adjusted to any desired value, without changing polymer concentration, pH, salt concentration or volume ratio. The modified PEG used in this way is either the negatively charged PEG-sulphonate, S-PEG,

$^{-}O_3S \cdot CH_2CH_2O(CH_2CH_2O)_xCH_2CH_2 \cdot SO_3^{-}$, or the positively charged

trimethylamino-PEG, TMA-PEG, $(CH_3)_3\overset{+}{N} \cdot CH_2CH_2O(CH_2CH_2O)_xCH_2CH_2 \cdot \overset{+}{N}(CH_3)_3$.

APPARATUS AND METHODS

a. Synthesis of TMA-PEG and S-PEG

The substituted PEG is prepared according to the reaction scheme shown in Fig. 1.

1. Bromo-PEG

Carbowax 6000 (500 g) is dissolved in 4 l of toluene, and 650 ml toluene is distilled off to remove traces of moisture. Then 55 ml of water-free triethylamine, distilled over phthalic anhydride is added. When the temperature drops to 35°, 25 ml of thionyl bromide (Fluka) dissolved in 100 ml toluene is added during 1 h with continuous stirring. Atmospheric moisture is excluded. The mixture is gently refluxed for 15 min. The hot liquid is filtered under suction, first through a sintered glass filter and then

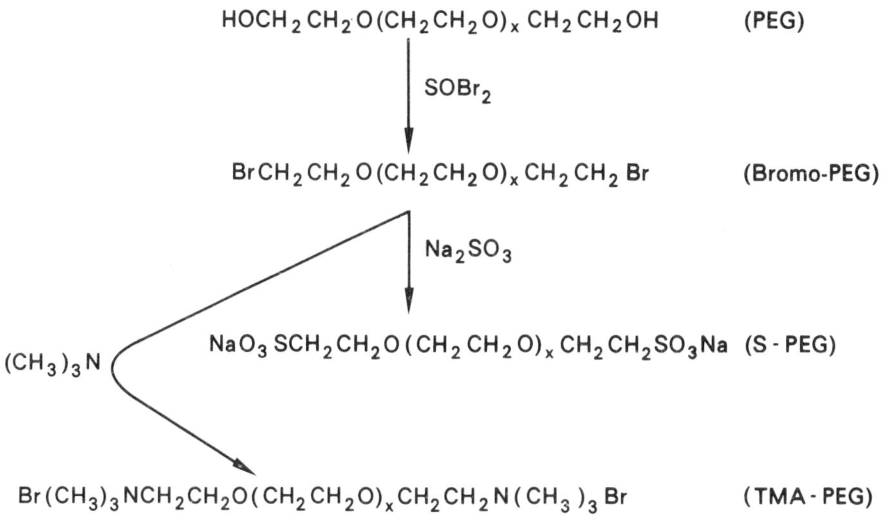

Fig. 1. Scheme for preparing the substituted PEG.

through 40 g of active carbon (Norit) that is layered on a filter paper in a 20 cm Buchner funnel. The filtrate is cooled down to 3° and the light-brown, greasy precipitate is collected by suction filtration in two 20 cm Buchner funnels. The wet precipitate is dissolved in 2.3 l of hot absolute ethanol and filtered through 40 g of active carbon when still warm. The polymer precipitates from the filtrate on standing at 3° (14 h). The precipitate is collected as before and washed on the filter with 3 l of ice-cold absolute ethanol. The polymer is dissolved in 3 l of absolute ethanol, filtered through carbon, precipitated and washed in the same way. If Carbowax 4000 is used the amounts of trimethylamine and thionyl bromide should be 110 and 50 ml respectively.

2. TMA-PEG

Bromo-PEG (300 g), M_n = 6,000, is dissolved in 2 l of absolute ethanol, and 25 g of trimethylamine (50 g for Bromo-PEG of M_n = 4,000) dissolved in

500 ml absolute ethanol (-20°) is added. The reaction is complete after 38 h on a 80° water bath. The solution is filtered first through 40 g of active carbon and then through additional 80 g. The solution is layered in 8 flat glass dishes (diam. 20 cm) and placed over conc. sulphuric acid in dessicators, which are evacuated. On standing overnight in the dessicators part of the polymer precipitates and the smell of trimethylamine disappears The mixture is warmed to get a solution and transferred to a round-bottomed flask. Ethanol is removed by evaporation in vacuum so that the final volume becomes 2.5 l. Freshly distilled ether, 1.5 l, is added and th polymer allowed to precipitate at 3° overnight. The polymer is collected on filter paper by suction and washed with 3.6 l of freshly distilled ether The wet filter is transferred to a round-bottomed flask and all solvent is evaporated in vacuum at 80° over a water-bath. When gas bubbles no longer are formed, the molten polymer is cast on a flat enamel dish.

3. S-PEG

Bromo-PEG (250 g) and anhydrous sodium sulphite (50 g) are dissolved in 1 l of water. Then 0.35 l ethanol is added and the solution is placed on a water bath at 80° for 15 h, whereafter 20 g of active carbon is added and the mixture is shaken for 2 h. After filtration the solvent is evaporated under vacuum at 80° and the polymer is dissolved in 1 l absolute ethanol. The undissolved salts are removed by filtration, and the polymer precipitates in the cold (-30°). The precipitate is collected by centrifugation (10 min at 11 000 *g*) and dissolved in 1 l of xylene. Traces of insoluble salts are filtered off, and the polymer precipitated at -30° by adding 0.5 l of ether. The precipitate is collected by centrifugation, washed wit ether, and recrystallized from 1.5 l of warm absolute ethanol.

b. Composition of a suitable biphasic system

To find a system in which the particles partition with the desired distribution ratio, G, the scheme given below may be followed.

1. The per cent proportions of dextran (M_n = 500,000) and PEG (M_n = 4,000 or 6,000) are chosen so that the system has a moderate settling time and so that the phases are not too viscous. High viscosity may make the mixing difficult. Suitable system compositions can be found in Albertsson's monograph (1). Note that the phase composition and the viscosity of the phases are temperature-dependent.

2. The pH chosen should be suitable for the particles in question, and the buffer used should have high buffer capacity at this pH. The concentration of buffer must be kept low (10 mM or less), lest the effect of the charged PEG be diminished by electrostatic shielding. Sucrose or sorbitol may be included to keep the system isotonic with respect to the particles.

3. The particles are partitioned in this dextran-PEG system. If they are found in the upper phase, G can usually be decreased by replacing part of

the PEG with S-PEG. G is here defined as:

$$G = \frac{\text{mass of material in the upper phase}}{\text{mass of material in the lower phase and on the interface}}$$

If the particles go to the interface or to the lower phase, G may be increased by using TMA-PEG. A series of phase systems is prepared where the proportion of PEG replaced by S-PEG or TMA-PEG is varied and the per-cent of the particles in the upper phase is determined. Fig. 2 illustrates a typical diagram obtained in such experiments. (Compare with the experimental curve given in Fig. 3.)

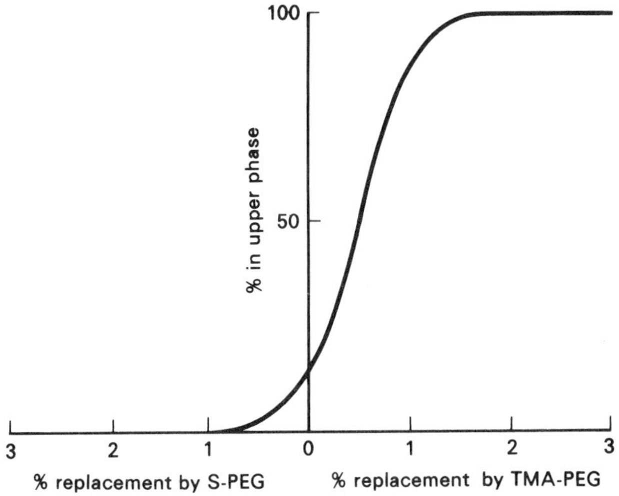

Fig. 2. Example of a phase diagram. The proportion of particles in the upper phase is shown in relation to the proportion of the PEG replaced by S-PEG or TMA-PEG

The per cent of charged PEG required for 50% of the material to be in the upper phase is determined from the diagram (G=1). This corresponds to 0.5% TMA-PEG in Fig. 2. A CCD is then carried out with the requisite system keeping the interface stationary. A thin-layer CCD apparatus is to be preferred (3), but even a CCD apparatus according to Craig gives good results, if only the settling time is not too brief. A stationary interface is preferable to a mobile one, since some of the material on the interface may drop down into the lower phase.

If the material is a mixture of two discrete types of particles, the extraction curve may have an inflection point as in the case of Chlorella (Fig. 3). In such a case it is more favorable to adjust the amount of charged PEG to the value of this inflection (1.4% TMA-PEG in Fig. 3). A CCD carried out under these conditions is shown in Fig. 4.

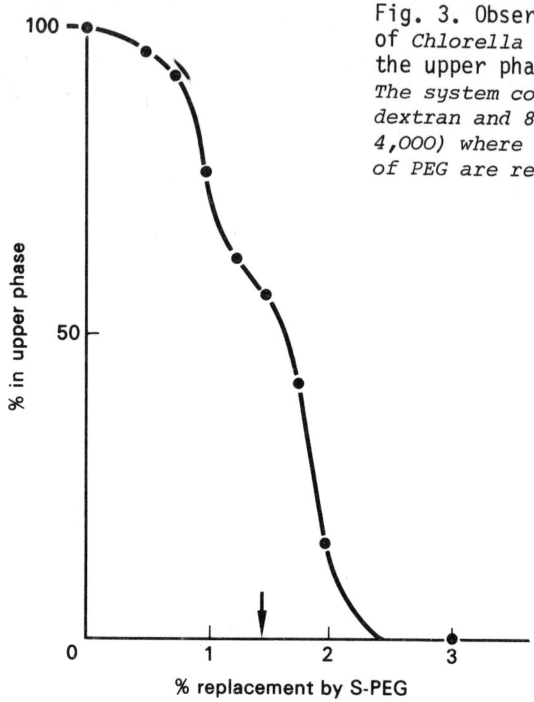

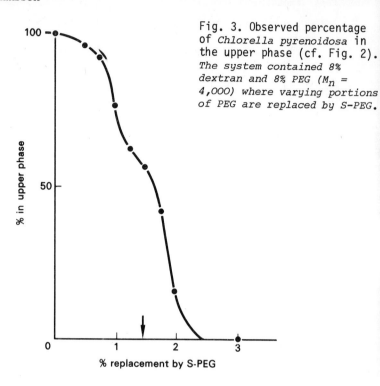

Fig. 3. Observed percentage of *Chlorella pyrenoidosa* in the upper phase (cf. Fig. 2). *The system contained 8% dextran and 8% PEG (M_n = 4,000) where varying portions of PEG are replaced by S-PEG.*

DISCUSSION

CCD is not the only way to utilize these biphasic systems. If only one of the two components of the Chlorella is to be purified, the best way to do so is to dissolve the material in the required phase and then take away the unwanted material by repeated extractions with the complementary phase.

Another way to use these systems is to start with a high enough percentage of S-PEG so that all types of particles in the mixture stay in the lower phase. By stepwise change in the content of charged PEG in the upper phase, it is possible to resolve some mixtures by extracting the components one after the other from the lower phase. The stepwise extraction can be carried out by using upper phases from systems with a gradually decreasing titre of S-PEG followed by those with increasing titre TMA-PEG. Referring back to Fig. 3 and starting with 5% S-PEG, the first extraction should be done at 1.4% S-PEG, the second at 0% S-PEG. This is of course not

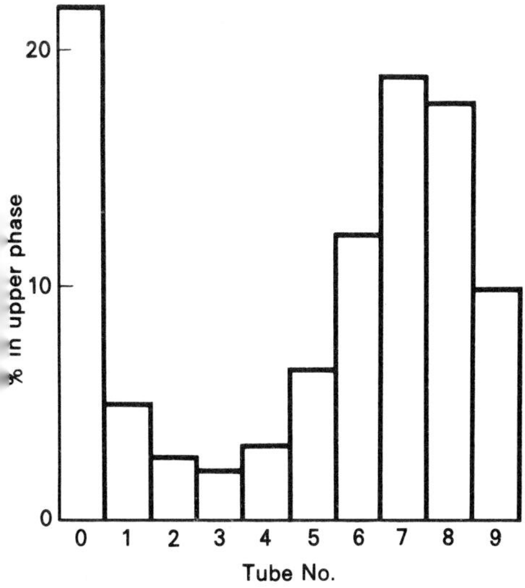

Fig. 4. An illustrative separation: CCD of *Chlorella Pyrenoidosa. The separation was performed at 21° in a system containing 8% dextran and 8% (PEG+S-PEG), M_n=4,000. Of the total PEG, 1.4% was in form of S-PEG. The composition can be written as 8% dextran, 7.89% PEG and 0.112% S-PEG. The system also contained 2mM potassium phosphate buffer, pH =6.8. Nine transfers were carried out using a Craig apparatus. Each chamber was filled with 1.5 g of the lower phase and 2.0 g of the upper phase. The material, about 0.5 mg, was present initially in chamber No. 0. Of the upper phase, 1.5 g (75%) was mobile (to assure a stationary interface. The two peaks would be somewhat better separated if the whole upper phase could be transferred. When the same experiment was performed by transfer of the entire upper phase manually by pasteur pipette, the two components become almost completely separated after only four transfers.*

practicable in this case when only two components are present.

Since the particles probably are separated mainly according to their electrical charge, the use of dextran-PEG systems containing charged PEG should be a complement to electrophoresis. Liquid-liquid extraction methods have, however, the advantage that they can easily be scaled up.

The use of charged PEG in dextran-PEG-water biphasic systems makes it possible to find quickly the most suitable partition ratio. The effect of charged PEG on the selectivity of the separation has not been studied for particles. Experiments with proteins have shown that the resolving properties of the systems are usually enhanced when S-PEG or TMA-PEG is used. The same may be true for particles, and such tendencies have been noticed in some preliminary experiments.

References

1. Albertsson, P.Å., *Partition of Cell Particles and Macromolecules*, 2nd ed., Almqvist & Wiksell, Stockholm; Wiley, New York (1971).

2. Albertsson, P.-Å., *Adv. in Prot. Chem., 24* (1970) 309.

3. Albertsson, P.-Å. in *Methods in Microbiology, Vol 5B,* eds. Norris, J.R. and Ribbons, D.W., Academic Press, New York, (1971), p. 385-423.

4. Walter, H., Garza, R. and Coyle, R.P., *Biochim. Biophys. Acta 156,* (1968) 409.

5. Johansson, G., *Biochim. Biophys. Acta 222* (1970) 381.

6. Karlstam, B. and Albertsson, P-Å., *Biochim. Biophys. Acta 216* (1970) 220.

7. Larson, C., Collin, C. and Albertsson, P.-Å., *Biochim. Biophys. Acta 245* (1971) 425.

8. Ericson, I., in *Exp. Cell Res.,* to be published.

9. Walter, H., *Progr. in Sep. and Purif., Vol II,* ed. Gerritsen, T., Wiley, New York, (1969) p.121.

10. Walter, H., Eriksson, G., Taube, O., and Albertsson, P.-Å., *Exp. Cell Res. 64* (1971) 486.

18 ZONAL CENTRIFUGATION*

George B. Cline
University College
University of Alabama in Birmingham
Birmingham,
Alabama 35294,
U.S.A.

Zonal centrifugation is a separatory technique which was introduced in principle by E.B. Harvey [1], who centrifugally separated most of the major organelles of sea urchin eggs within the confines of the egg itself. These were separations done in a density gradient of cytoplasm, and resulted in the gross separation of lipid, nucleus and mitochondria in the 'white half' and pigment in the 'red half'. We now utilize a variety of solutes to form special density gradients in which we separate an extremely wide variety of particulates. These particulates range in size from serum proteins [2] to whole animals [3]. This article reviews the elementary principles of zonal centrifugation, presents lists of particulates already effectively separated, and discusses general considerations in choosing gradient materials and gradient shapes. Whilst a variety of zonal centrifuge rotor systems are dealt with in the discussion, the principles and considerations still apply to gradient separations in centrifuge tubes. Finally, selected separations from the author's laboratory are presented.

The term 'zonal centrifugation' implies that zones of material are separated by centrifugation. In the simplest case, a mixture of particulates is layered on the top of a density or viscosity gradient of some appropriate solute, and gravitational force is used to drive the particulates through the gradient. The gradient is used to keep the separating particulates in zones. If the gravitational force and time factors are sufficient, particulates may be sedimented through the gradient to where the density of the particulate actually equals the density of the gradient at that point. This type of separation is called an isopycnic separation and is different in principle from a second case where the separation (centrifuge) is stopped and the fractions recovered before particles of a particular class reach their isopycnic level. This second procedure is called a rate separation, and the position of the particulates in the gradient is most directly related to the size of the patticulate. Either isopycnic or rate separations can be made easily in swinging bucket, angle head or zonal rotors.

* *This article represents a general account of the 'state of the art', as more systematically considered in Vols. 1 and 3 of this series.* - Editor.

Special zonal centrifuge rotor systems have been designed to permit replacing the sample on top of a density gradient (4,5). These so-called continuous-sample-flow-with-isopycnic-banding ('flo-band') rotors are of the utmost value to those interested in isolating small amounts of particulates from large volumes of sample. These zonal rotors are among the latest designed by Dr. Norman G. Anderson of the Molecular Anatomy Program of of the Oak Ridge National Laboratory. Data are presented below to show the utility of the latest high speed J-I system in the isolation of selected serum proteins.*

TABLE 1. Reported separations

Organisms

- Fish eggs
- Fish larvae
- Phyto- and zoo-plankton

Subcellular particulates

- Nuclei, mitochondria, chloroplasts
- Plasma membranes, smooth and rough endoplasmic reticulum
- Peroxisomes (plant and animals)
- Lysosomes
- Ribosomes, ribosomal subunits, polysomes
- Synaptosomes

Viruses

- Australia antigen
- Infectious hepatitis (Marmosets)
- Herpes simplex
- Burkitt lymphoma
- Adenoviruses
- Rhino viruses
- Myxoviruses
- Fox (insects)
- Granulosis (insects)
- Nuclear and cytoplasmic (insects)

Cells

- Tissue culture cells
- Red blood cells
- White blood cells
- Yeast
- Bacteria - *Bacillus, E. coli*
- Blue green algae
- Trachoma agent

Macromolecules

- Serum proteins, 27S, 19S, γ, AHF.
- Serum lipoproteins
- DNA, RNA
- Enzymes
- Collagen
- Fibrin
- Insulin inhibitor
- Thyroglobulin
- Viral antigens (components)

Miscellaneous particles

- Sands, silts
- Clays
- Mineral colloids

SEPARATIONS ACHIEVABLE

In cataloguing the variety of particulates separated in gradients it is apparent that an abbreviated but representative list may be organized in several ways, one being the scheme in Table 1. Each investigator has had to consider the possible effects of the gradient material on the particle he was isolating. The most important factors are listed in Table 2.

* *For these and certain other separations (fish eggs and larvae, insect viruses), method descriptions are given in Vol 3 (viz. the book cited in refs. 2 & 3).*

TABLE 2 Factors affecting the choice of gradient material

1. Does the gradient material give the separation desired?
2. Will the material allow of adequate resolution?
3. Is the material toxic to the particulate(s) or activity to be measured?
4. Can the material be separated from the particulates?
5. Will the contaminating material affect subsequent assays?
6. Need the material be recovered after use for reuse?
7. Will the material permit detection of the zones by light absorption?
8. Will the material corrode the zonal rotors or other equipment?

In considering the above list of factors it is important to know the most generally available types of solutes for use in making liquid gradients. More extensive lists have been published elsewhere (6). The following list outlines the most popular solutes, most of which may be used to make either hot or cold gradients:-

Sucrose	Ficoll
Caesium chloride	Albumin
Dextran	Sorbitol

METHODS

Sample preparation

The preparation of starting samples for zonal centrifugation in zonal rotors is quite similar to preparation for separation in ordinary centrifuge tubes. The one difference is in quantity, and the amount of sample needed depends upon the type of zonal rotor to be used. For example, the B-XIV rotor gives best separation when a 10 to 25 ml starting sample is used. The larger B-XV zonal rotor will conveniently use from 20 to about 60 ml of starting sample. Of course, the size depends upon the type of separation desired and whether the sample is put on top of the gradient or on the bottom of the gradient and floated up.

The flo-band rotors can utilize very large sample sizes since the sample is made to flow through the rotor continuously. For a collection of such components as cellular membranes or mitochondria, we usually make from 5 to 20 l of a 5% or 10% homogenate. The possible sample size under the stated conditions is thus 2 kg. The most important point to be made about sample size in the flo-band types of rotors is not how much sample you can put through the rotor, but how much separated component can be held with good resolution in the gradient you are using. While the marginal flow-through volumes with large-volume rotors such as the K-II or K-X are above 200 l, it is frequently difficult to recover the product from the rotor if the product was in high concentration in the flow-through starting sample.

Types of gradient

A density gradient is used primarily to stabilize the zones of sedimenting particulates and keep them separated. Since most zonal rotors are basicall hollow bowls with special cores to section off the internal volume, each section turns out to be a sector-shaped compartment. Radial dilution of sample thus becomes one of the biggest factors in many separations where the original concentration is low. The shape of the gradient may thus aid the separation immeasurably.

Since many zonal rotor users use a commercial type of gradient-forming device, the easiest gradient to make is one in which the density increases *linearly with the volume*. Due to the radius-to-volume relationships in the rotors this turns out to be a convex gradient when in the rotor. This is not the most desirable shape of gradient for most separations, since the gradient capacity (7) is lowest at the top of the gradient where you need capacity the most and is the highest deeper into the rotor where you need it the least. The one advantage of this shape of gradient is the zone-narrowing capabilities near the bottom of the gradient.

A second type of gradient is *linear with rotor radius*. This gradient has constant capacity throughout and is one of the most commonly used shapes, since it lends itself to easy calculation of where the particulates are at any given time. If the steepness of the gradient is right, the particles may travel at a constant rate (isokinetic) because the gravitational force is offset by the increased drag (resistance) on the sedimenting particulate.

Gradients may also be *convex* in the rotor. This shape of gradient gives a high capacity at the top to support the starting sample but then decreases in capacity toward the periphery of the rotor. Since any zone of sedimenting particulates is being radially diluted the capacity of the gradient is generally adequate throughout the sedimenting distance (unicapacity gradient).

Gradients may also be designed which keep a nearly constant zone size. These *isovolumetric* gradients (8) promise increased resolution.

Since the capacity of a gradient to hold sample is related to the steepness of the gradient, *discontinuous gradients* have perhaps offered the highest resolution possible. Discontinuous gradients are made from stacking in a series of discrete density solutions and letting diffusion set up a steep gradient between any two solutions. Merely stacking solutions will give several zones of mixed particles and little separation will be achieved. For maximal effectiveness, the sample should first be fractionated on a gradient linear with radius or linear with volume. The positions of the zones then dictate the volumes and numbers of solutions to be used in making the discontinuous gradient. The gradient is arranged so that under the conditions of centrifugation each zone ends up

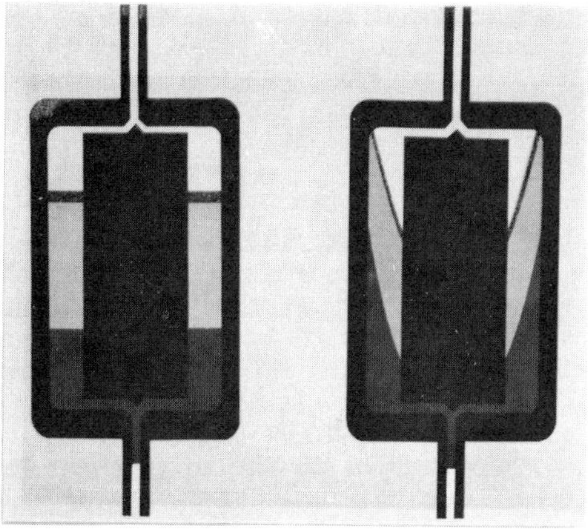

Fig. 1. A) The configuration of a continuous-sample-flow-with-isopycnic-banding rotor (flo-band) when the rotor is to be used for a rate separation (non-flow-through). A buffer, sample and gradient are pumped into the rotor through the bottom line. B) The rotor with sample and gradient is accelerated from rest. The gradient orients on the rotor wall.

Fig. 2. *(above)*. When the rotor is at speed, the sample and gradient is accelerated from rest. The gradient orients on the rotor wall.

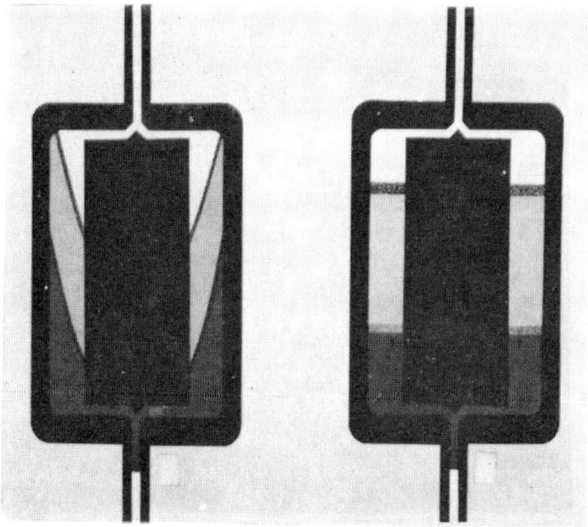

Fig. 3 *(left)*. A) When the rotor is decelerated the gradient begins to reorient back to its position at rest. B) When the rotor stops the gradient is in the original position. The gradient is recovered by draining out the bottom of the rotor.

on one of the steep diffusion interfaces. This concentrates the material in the zone and increases resolution over other types of gradients. However, if the duration of centrifugation is quite long, diffusion will smooth out the gradient and the resolution approximates that obtained if one starts with a smooth linear gradient. One should not make the mistake of starting with an uncritically chosen discontinuous gradient and expecting the particulates seen on the interface to be those desired. Such zones may well be mixtures of particulates which did not get either the proper gradient shape or the proper time-gravity considerations to separate adequately.

Perhaps the highest resolution attainable in theory is with the decreasing viscosity gradient where as the density increases with rotor radius, the viscosity decreases. These gradients must be made up with two or more solutes with the highest viscosity solute at the top of the gradient. As the largest particulates sediment into the gradient they 'see' less drag and higher centrifugal force. They thus accelerate away from the slower moving particulates and high resolution is attained.

Gradients may be made also with variables in the solvent material besides the solute. Such variables include pH, ionic strength, organic solvents, heavy metals, enzymes, ammonium sulphate, polyethylene glycol, and other materials that may be put into gradients as zones through which particulates must sediment. It is thus possible to design and achieve some rather sophisticated separations-extractions-precipitations during a single zonal separation.

APPLICATIONS

Since some of the applications illustrated below involved flo-band zonal rotors, attention should be drawn to the simplicity of achieving separations in these rotor types. Fig. 1 shows a cross-sectional diagram of a flo-band rotor (could be any of the Model K, RK or J-series rotors). The rotor is filled with the gradient material while the rotor is at rest. This is usually accomplished by pumping two or three chosen solutions into the bottom of the rotor. If a rate separation is made, the sample is pumped in between the overlay and the top of the gradient. Fig. 1B shows the acceleration phase for a rate separation. The gradient and sample climb the walls of the rotor and orient on the rotor wall. Fig. 2 shows the orientation of solutions at rotor speed. The separation is achieved at this stage. Fig. 3A shows deceleration, where the gradient and separated sample reorient off the rotor wall and come back to rest when the rotor stops (Fig. 3B). The fractionated sample is collected from the rotor by letting it drain out of the bottom.

Fig. 4 shows combined data from two separations in the J-I flo-band rotor, with human serum (dotted line) and Factor VIII (antihaemophilic factor) fractionated by polyethylene glycol (PEG). These separations each took 70 min at a rotor speed of 55,000 rev/min using sucrose gradients at 30°. This is perhaps the first time that AHF activity has been shown to

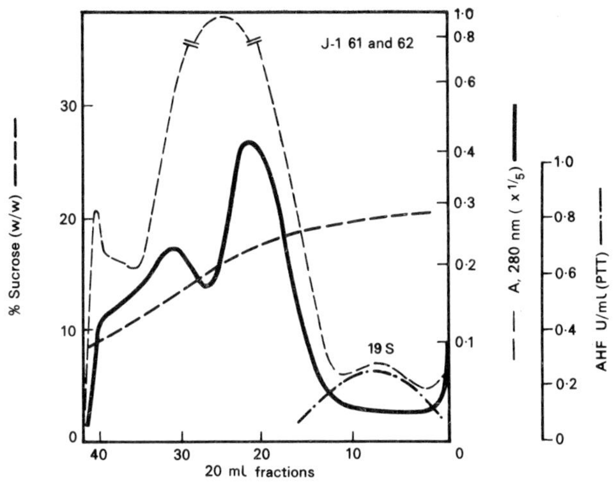

Fig. 4. A composite of two separations of serum proteins done in the J-1 flo-band rotor at 55,000 rev/min in 70 min using sucrose gradients at 30°. *The 19 S macroglobulins and antihaemophilic factor are separated from the other serum proteins.*

have a sedimentation value similar to macroglobulins. It is not yet known whether AHF is carried by the α_2 macroglobulins or is separate and distinct from it.

Figure 5 shows a spinach chloroplast separation in the J-I rotor. The main zone of chloroplasts is centred at about 40% sucrose (w/w). Another zone is centred at about 51%. The starting sample was a cheese-cloth filtered homogenate suspended in 3 litres of 17% sucrose. The flow-through rate was 4 l/h at a rotor speed of 4,000 rev/min.

Figure 6 shows a similar preparation of spinach chloroplasts separated on a discontinuous gradient in the batch-type B-XIV zonal rotor. The two main zones of chloroplasts were centred at 48% and 39% sucrose. The large zone to the right in the figure is primarily soluble material which did not sediment into the gradient in 35 min at 30,000 rev/min.

Figure 7 shows a separation of the blue-green algae, *Microcystis aeruginosa*, in the B-XIV zonal rotor. Cells which contained gas vacuoles were lighter in density than other cells and so floated to the top of the gradient (Fractions 1-3).

Figure 8 shows a combined rate and isopycnic run on a mixture of

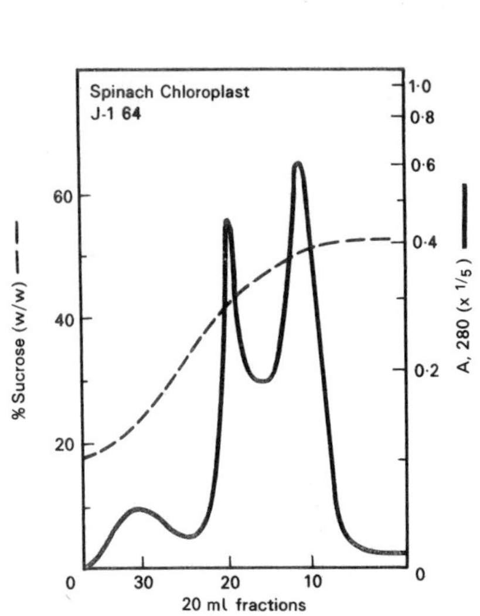

Fig. 5. Continuous flow collection of chloroplasts from spinach leaf homogenate in the J-I rotor using sucrose density gradient.

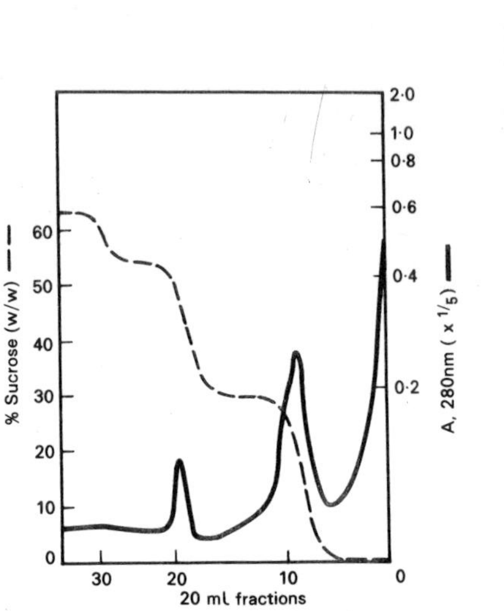

Fig. 7. A separation of gas vacuole-containing blue-green algae in the B-XIV zonal rotor. See text for details.

Fig. 8. Bacteria separation in sucrose gradient in B-XIV zonal rotor.

Fig. 6 *(below)*. A combined rate and isopycnic separation of gluteraldehyde-fixed chloroplasts in the B-XIV zonal rotor.

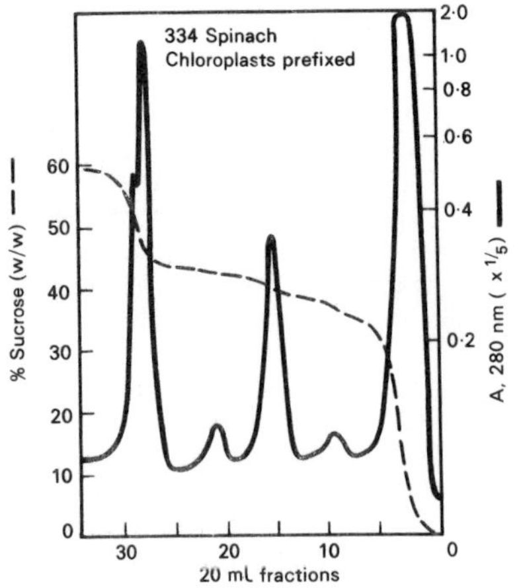

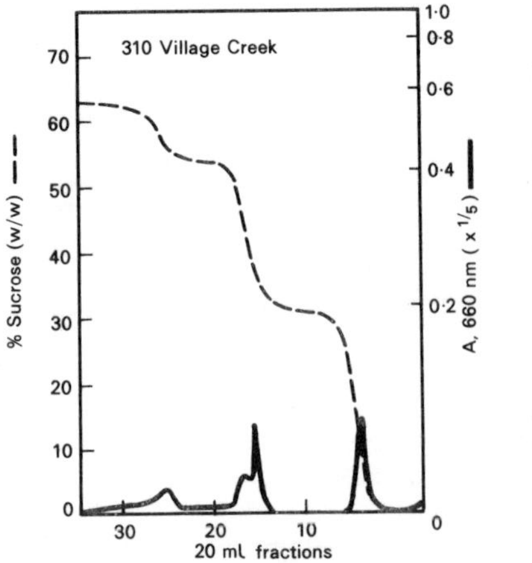

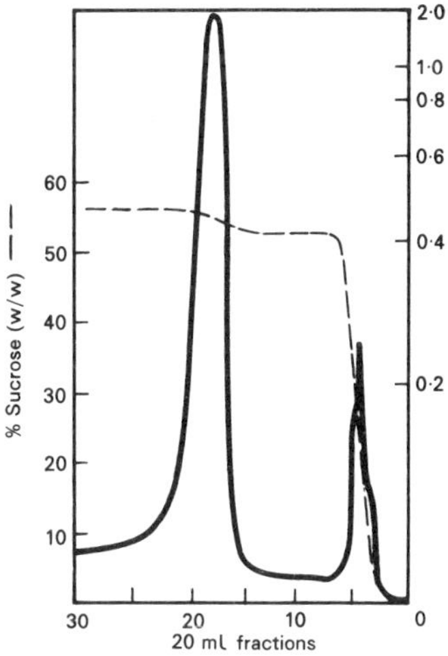

Fig. 9. Isolation and purification of the nuclear polyhedrosis virus inclusion bodies of *Heliothis zea* on high resolution discontinuous sucrose gradient.

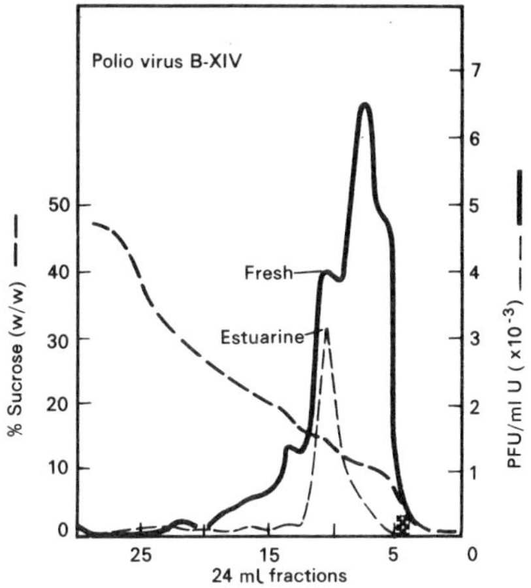

Fig. 11. A composite of two separations of polio virus samples incubated in either fresh or estuarine water. B-XIV rotor and sucrose gradient for 3 h at 30,000 rev/min.

Fig. 10 *(below)* A quasi-isopycnic separation of an insect pox virus centred at about 62% sucrose in the B-XIV zonal rotor.

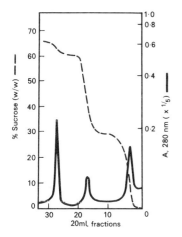

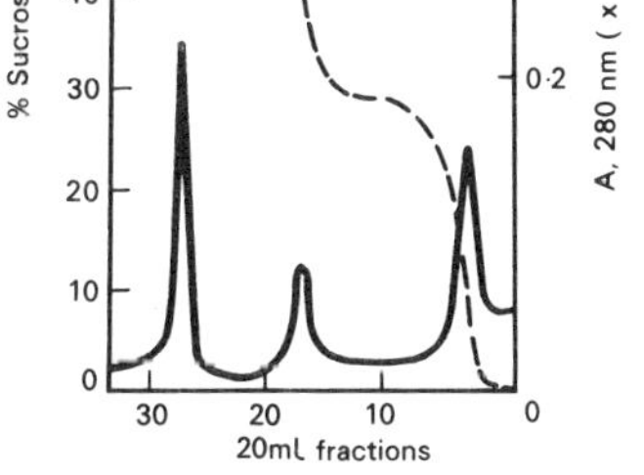

Fig. 12 *(below)* An RK-II flo-band rotor separation (7 l/h at 30,000 rev/min) of the nuclear polyhedrosis virus inclusion bodies of a putrefied *Heliothesis zea* sample. The virus is centred at about fraction 13.

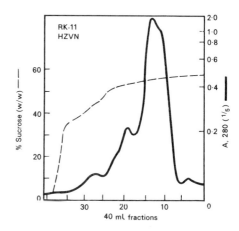

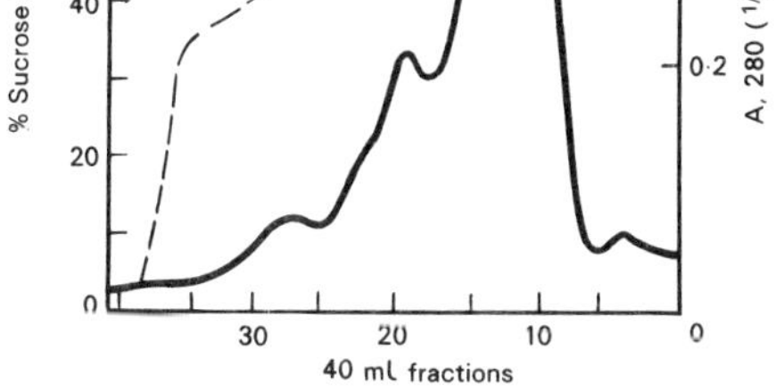

bacteria collected from natural water. All three zones contained coliform bacteria. The centre zone shows the effect of particulates 'breaking through' a steep interface which was set up to retard slower moving cells. Both *Bacillus* and *E. coli* will band quasi-isopycnically at about 54% w/w) sucrose.

Fig. 9 is a B-XIV zonal rotor separation of the nuclear polyhedrosis virus of the cotton boll worm, *Heliothis zea*, from contaminating bacteria. The virus inclusion bodies band isopycnically at about 56% sucrose while the bacteria will band isopycnically at about 54% sucrose. The discontinuous gradient used in this separation was effective in keeping the bacteria back at the starting zone (Fraction 5).

Insect pox virus was separated in Fig. 10 on a discontinuous gradient in the B-XIV rotor. The pox virus can be slowed down and recovered from 60-62% sucrose but longer sedimentation will drive it slowly through even 66% sucrose. Bacterial contaminants were found in the two other zones to the right.

Density gradients are quite useful for isolating aggregated particulates. Figure 11 shows two separations of polio virus plotted together. Sedimentation time was 3 h at 30,000 rev/min, and the starting zone is denoted by the small marker at fraction 5. Polio virus appears to aggregate more in estuarine water than in fresh water. The starting discontinuous density gradient diffused into a nearly linear gradient over much of the volume of the rotor.

One of the flo-band rotors, the RK-II, was used to collect the nuclear-polyhedrosis virus of *Heliothis zea* from a putrefying culture (Fig. 12). The inclusion bodies (which contain the virions) banded isopycnically at about 56-57% sucrose but had a low-level contamination of bacterial particles from the zone centred at about fraction 20.

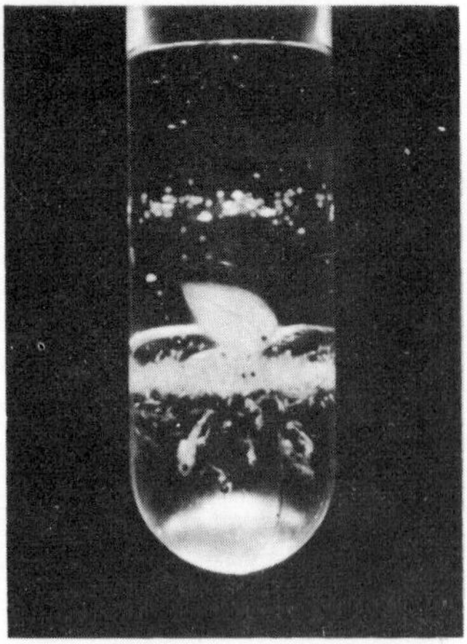

Fig. 13. A photograph of a swinging bucket centrifuge tube in which a plankton tow net sample was fractionated in a caesium chloride gradient. See text for details.

Fig. 13 shows an interesting separation of material collected in a marine plankton tow. The collected material was layered over a caesium chloride density gradient which ranged from about 1.1 at the top to 1.45 at the bottom. The photograph shows a prominent zone of fish eggs at the top centred at about density 1.22 to 1.25, some fish larvae at about 1.30 to 1.32, most copepods at 1.35 to 1.36 and other fish larvae at about 1.38 to 1.40. This rapid plankton-sorting method can use either swinging bucket, angle head, or zonal rotor systems. For small samples 20 min at 2,000 rev/min in a refrigerated laboratory centrifuge (1200 to 1500 *g*) is quite convenient.

Acknowledgement

The author gratefully acknowledges the expert assistance of Martha K. Dagg and the support of Electro-Nucleonics, Inc.

References

1. Harvey, E.B., *Biol. Bull. 69* (1935) 287.
2. Cline, G.B., Dagg, M.K., Srour, J., Wickerhauser, M. and James, H. in *Methodological Developments in Biochemistry, Vol. 3.-Advances with Zonal Rotors* (E. Reid, ed.), Longman, London (1973) *in press*.
3. Cline, G.B. and Doggett, L.F. in *Methodological Developments in Biochemistry, Vol. 3.- Advances with Zonal Rotors* (E. Reid, ed.), Longman, London (1973) *in press*.
4. Anderson, N.G., Barringer, H.P., Amburgey, Jr., J.W., Cline, G.B., Nunley, C.E. and Berman, A.S. *Natl. Cancer Inst. Monograph 21* (1966) 199.
5. Cline, G.B., Anderson, N.G. and Fennell, R. in *Separations with Zonal Rotors* (E. Reid, ed.), Wolfson Bioanalytical Centre, University of Surrey, Guildford (1971) p. A-3.1.
6. Cline, G.B. and Ryel, R.B. in *Methods in Enzymology* (S.P. Colowick and N.O. Kaplan, eds.) *22* (1970) 168.
7. Spragg, S.P. and Rankin, C.T., *Biochim, Biophys. Acta 141* (1967) 164.
8. Pollack, M.S. and Price, C.A., *Anal. Biochem. 42* (1971) 38.

19 SEPARATION TECHNIQUES FOR HAEMOPOIETIC CELLS BASED ON DIFFERENCES IN VOLUME AND DENSITY

K.A. Dicke
Radiobiological Institute
Rijswijk
The Netherlands

Methods of fractionation of mammalian cell suspensions fall into two main categories: those which make use of differences in functional properties of the cell population and those which are based on differences in physical characteristics of the cells. Adherence column separation is a classic example of fractionation by function since this method is based on the ability of phagocytosis of certain cell populations. Methods based on differences in physical characteristics depend on density, size and electrophoretic properties. Density gradient centrifugation is widely used, and size-based separation has also been achieved with bone marrow cells from different species. The aim of fractionation of a complex population of cells is either analytical, that is to study the properties of a particular sub-population of cells under varying conditions, or preparative, that is to obtain a purified sub-population of cells to be used in further experiments. Our discontinuous density gradient technique is preparative, to separate bone marrow cell suspensions in rodents and primates in order to obtain fractions poor in lymphocytes and rich in stem cells, suitable for transplantation.

For many problems in cell biology it would be advantageous if one could separate a heterogeneous cell suspension into various fractions which are homogeneous in function. There are often two experimental objectives which may be mutually exclusive in cell separation experiments. The first objective is the analysis of a cell population. This usually involves continuous distributions, rather than stepwise distributions, and a careful analysis of the fractions. Small numbers of cells can yield useful results with such an analytical experiment. The second objective is to separate cell types as a preliminary step of an experiment, a large number of cells being required. One of the most evident applications of a preparative procedure is in the field of bone marrow transplantation. This is the prevention of acute Graft *versus* Host (GvH) disease by means of selective elimination of the lymphocytes from the graft. Lymphocytes are responsible for this syndrome of inducing an immunological reaction against the host. Generally the acute GvH disease leads to death of the recipient 7-14 days after allogeneic bone marrow transplantation (1). Experimental evidence is available in mice that the severity of the acute GvH disease is closely related to the number of lymphocytes present in the graft (2): the grafting

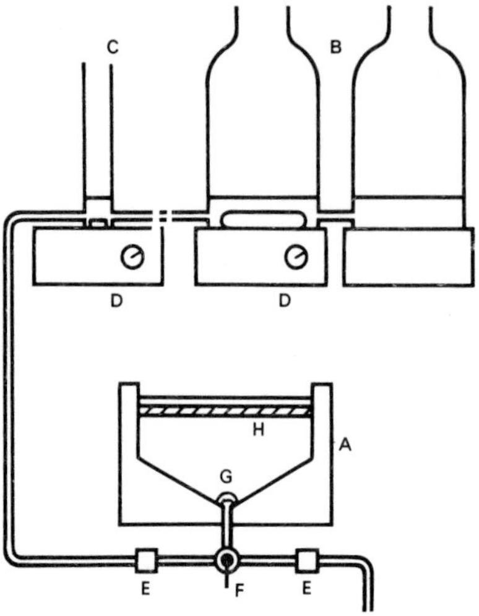

Fig. 1. Scheme of the velocity sedimentation technique of Miller and Phillips (6). A - sedimentation chamber; B - gradient maker; C - vessel for cell suspension; D - magnetic stirrers; E - valves for regulation of the flow rate; F - three-way valve; G - baffle for dissipation of the upward momentum of the fluid as it enters the chamber; H - cell band shortly after loading has been completed.

of smaller numbers of cells decreases the risk of acute GvH disease. The elimination of the lymphocytes from the marrow graft should be very selective since the haemopoietic stem cell (HSC) population, comprising only a very small part of the bone marrow cell population (approx. 1 in 1000), has to be grafted. The HSC is the only cell type in the marrow capable of restoring or replacing the haemopoietic tissue of the recipient, which is usually the aim of bone marrow transplantation (3). Therefore the preparative objective is to obtain fractions containing as many HSC and as few lymphocytes as possible.

Analytical experiments and preparation procedures often have different requirements; analysis of a cell population necessitates a great number of fractions (30-50) whereas in the preparative experiment 3-5 fractions might be sufficient. From these points of view fractionation techniques are reviewed which are in an advanced state of development and which are based on differences in cell volume and density.

CELL SEPARATION BASED ON VOLUME

The only technique for separating cells by volume which is in a highly advanced state of development is the velocity sedimentation technique as described by Miller and Phillips in the Ontario Cancer Institute, Toronto (4-6). This technique has been schematically depicted in Fig. 1. The cells are placed in vessel C from which they are loaded into the sedimentation chamber (A). In order to obtain an optimal separation the cells should start from a thin band layered on top of the fluid in the chamber.

Table 1. Sedimentation properties of different cell populations in mice from R.G. Miller & R.A. Phillips (6)

Cell	s value, *mm/h*
erythrocytes	2.0 ± 0.1
lymphocytes	2.9 ± 0.2
rosette-forming cells	3.0 ± 0.3
CFU *[defined below]*	3.9 ± 0.2
granulocytes	4.3 ± 0.2
7S - plasma cells	4.4 ± 0.3
19S - plasma cells	4.7 ± 0.2
mitotic mouse L cells (C3H origin)	16.3 ± 0.3

In order to allow a layer of cells to form initially on top, a shallow gradient is required, which is located under the thin layer of cells and consists of 0.2 to 2% albumin. This gradient is prepared by running 2% albumin together with a buffered solution into the velocity chamber, the solutions coming out of their respective reservoirs (B). Once the cells and the gradient are present in the chamber, the velocity sedimentation bottle is kept on a vibration-free bench for 3-5 h during which separation will take place. Then the fractions are collected through the bottom (f). In this system the Toronto group concluded that the sedimentation rate (s) of the cells is proportional to the square of their radius (r) according to the following formula: $s = kr^2$ where k represents all the constants of the system. It is evident that the smaller the volume of the cells, the slower their sedimentation rate.

This method of separation is used primarily as an analytical method. Sedimentation rates of various cell types can be measured as is demonstrated in Table 1. 'CFU' (Table 1) stands for 'colony-forming unit, spleen', the cell which is able to form a colony in the spleen of a lethally irradiated mouse recipient. Such a colony consists of granulocytes and erythrocytes and their respective precursors, as well as megakaryocytes. This cell type is considered to be the HSC* and is present in the spleen and bone marrow of the mouse (7,8). Evidently its velocity sedimentation rate is low and close to that of the lymphocyte population (Table 1).

This method has also been used for preparative purposes: e.g. prevention of GvH† in clinical allogeneic bone marrow transplantation. However, its application seems limited since very large chambers have to be used in order to separate large quantities of cells. Moreover, the fractionation procedure for clinical transplantation takes too long (24-36 h), which may influence the viability of the HSC.

* *defined opposite* † *defined on p. 175*

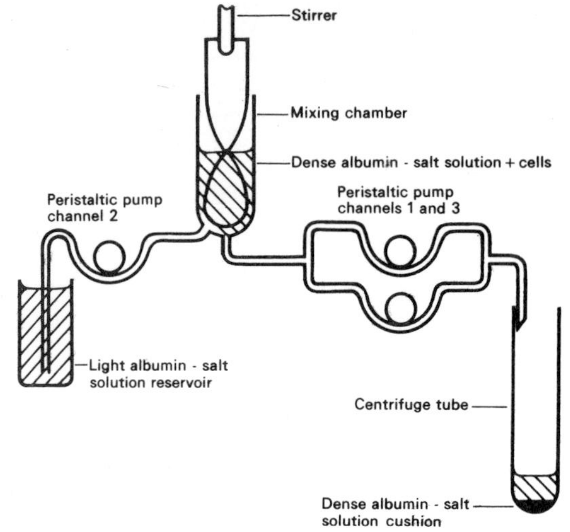

Fig. 2. Scheme of Shortman's continuous albumin density gradient gradient centrifugation technique (12).

Fig. 3 *(below)* Scheme of discontinuous density gradient centrifugation technique (17).

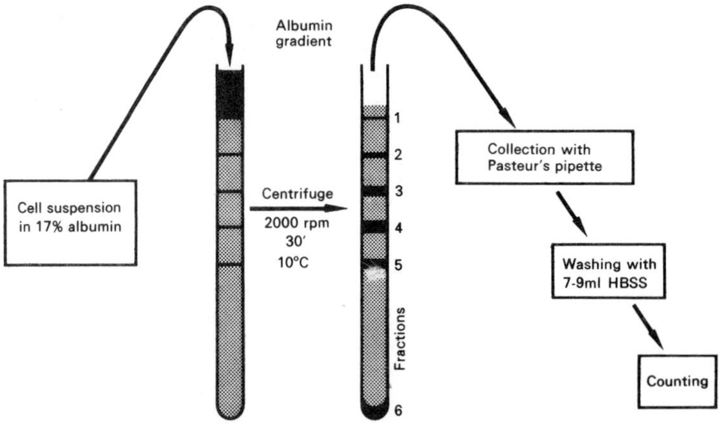

CELL SEPARATION BASED ON DENSITY

Separation based on differences in density was first described by Ferrebee (9) who tried to concentrate red cells which contained malaria parasites. Leif and Vinograd in 1964 introduced a reproducible and sensitive technique for the separation and analysis by density of erythrocytes, involving centrifugation to equilibrium in continuous albumin gradients (10). A comparable technique to that of Leif has been developed by Shortman (11,12) which is the most reproducible analytical procedure so far and is depicted in Fig. 2. A pump with 3 channels is used, one channel to deliver the light albumin to the mixing chamber and two parallel channels to transfer the developing gradient to the centrifugation tube. The position of the inlet and outlet holes of the mixing chamber, the stirrer shape, stirrer position and stirring rate are all critical. In order to improve the resolution of the gradient, Shortman emphasized the importance of controlling the physical characteristics of the albumin solution (13) and of the technical procedure of collecting the cells after centrifugation. A more detailed description can be found elsewhere (12).

Shortman used the separation procedure for the analysis of the cells of the immune system, in particular the lymphocytes. With regard to the preparative use of the continuous gradient method, it can be employed as such if the density range in the gradient is increased. However, it still necessitates a fairly complex gradient-generating procedure, and a careful checking of the particular density containing a certain population of cells utilizing a rather complicated bromobenzene gradient.

The fractionation method of choice for preparative purposes, e.g. in the field of bone marrow transplation to prevent acute GvH in the grafted recipient, must fulfil 3 requirements:

1. effective elimination of lymphocytes;
2. minimal loss of HSC;
3. technical simplicity, to allow widespread use of this method in clinical bone marrow transplantation.

With this in mind we developed a discontinuous, stepwise albumin gradient technique which is schematically represented in Fig. 3. The gradient consists of layers of bovine serum albumin (BSA). Solutions of different concentrations are layered with the aid of a Pasteur pipette in a glass centrifugation tube in the following sequence: 27% albumin on the bottom of the tube; 25%; 23%; 21%; and 19%. The albumin solutions are prepared from a 35% albumin stock solution in tris buffer by dilution with various volumes of a sodium chloride phosphate buffer. A detailed description of the technique has been presented elsewhere (14-17).

After layering the various albumin solutions, the cells suspended in a 17% albumin solution are pipetted on top of the 19% albumin layer. The tube is centrifuged at 2,000 rev/min (1,000 *g* at the bottom) during 30 min, after which distinct layers of cells are visible in the gradient near the density interfaces. The cell fraction between the 17 and 19% BSA layers is labelled fraction 1, that between the 19 and 21% BSA layers fraction 2, and so on to fraction 6 at the bottom of the tube. After collecting and washing, the cells are resuspended in Hank's solution and counted. In order to obtain an optimal and reproducible separation several physical characteristics in the gradient had to be controlled, as listed in Table 2.

Table 2. Physical parameters of the gradient to be controlled in order to get optimal resolution

1. pH of the albumin stock solution - pH 5.1
2. Osmolarity of the albumin stock solution -

   mouse gradient : 330 mOsm; monkey gradient : 340 mOsm;
   human gradient : 360 mOsm
3. Albumin concentration of the different solutions in the gradient - The concentration of albumin in a solution is related to the density of that solution.
4. Temperature - Fixation during centrifugation at 10°.
5. Number of cells per gradient - Overloading the gradient causes aggregation of cells, which disturbs reproducible resolutions

Table 3
Efficiency of separation of GvH active cells and haemopoietic stem cells in rodent and primate haemopoietic suspensions by the discontinuous gradient technique

Optimal fraction	Lymphocyte conc. factor*	HSC conc. factor†	Recovery of HSC ẋ	Ratio HSC/lymphocyte ẋẋ
MOUSE spleen	1/10	10	15 - 25	100
MONKEY marrow	1/10	10	15 - 25	100
HUMAN marrow	1/20	15	30 - 40	300

* determined by the PHA response test. In the mouse also the Simonsen assay was used.

† determined by the agar colony formation. In the mouse also the spleen colony assay was used.

ẋ expressed as a percentage of the number of stem cells in the unfractionated suspension.

ẋẋ The ratio HSC/lymphocyte in the original suspension is set at 1.

It can be noted that for each species a different osmolarity of the stock solution (from which the gradient is prepared) had to be used. It was found that in monkeys and humans a better separation between lymphocytes and HSC was obtained using gradients with a higher osmolarity than is routinely employed for mouse suspensions (18,19). Determination of the optimal value for each of the various physical properties of the gradient system for different species could only be done with the aid of sensitive quantitative assays for lymphocytes and HSC. In mice, the Simonsen assay and the CFU-assay, both *in vivo* methods, are established quantitative assays for lymphocytes and HSC respectively (7,20). The application of these assays is limited to mice and cannot unfortunately be extended to the primate, for which only *in vitro* methods can be employed. Therefore new test systems for the two cell populations were introduced. For lymphocyte detection the PHA response test is used, based on the phenomenon that PHA (phytohaemagglutinin) can transform lymphocytes into blastoid cells which incorporate ^{3}H-thymidine (15). The thin-layer agar colony technique, based on the observations of Pluznik (21) and Bradley (22) that mouse bone marrow is able to form colonies in agar, appeared to be a reliable detection method for HSC after proof that the HSC does take part in the formation of colonies *in vitro* (23,24).

With the aid of these four tests the efficiency of separation of the two cell populations was determined; the results are summarized in Table 3. These data, published *in extenso* elsewhere (3), clearly demonstrate a high efficiency of the separation of lymphocytes from HSC. This is reflected in the high ratio obtained by dividing the HSC concentration by the lymphocyte concentration after marrow cell separation. The retrieval of HSC from the lymphocyte-poor fractions, however, is only moderate. With regard to clinical bone marrow transplantation, the implication is that sufficiently high numbers of 'purified' HSC for the grafting of *adult* recipients cannot always be obtained.

Although these data show a definite separation, they still do not answer the question of whether it is efficient enough to prevent acute GvH disease. This issue is the main criterion of separation efficiency. Indeed, it has been shown that small numbers of purified stem cells from monkey marrow restore haemopoiesis in lethally irradiated allogeneic monkey recipients without causing acute GvH (Table 4). This model resembles the situation in clinical bone marrow transplantation and is therefore the model of choice. The control group treated with unfractionated bone marrow dies from acute GvH within 13 days on the average after grafting (Table 4). In the group treated with the stem cell fraction the delayed form of GvH usually developed; some animals died from infections. It should be noted that random host-donor combinations were used. In humans, interpretation of the results of transplantation of bone marrow fractions is complicated by the fact that the majority of recipients were in the final stage of the disease and bone marrow transplantation was considered to be the last resort. Moreover, various other treatments were applied, directed at the prevention of acute GvH. Although the results are highly promising, a detailed description as given elsewhere (3) is beyond the scope of this paper.

This gradient has also been employed for analytical studies although in the literature the value of any discontinuous gradient for this purpose has been disputed. For this purpose the increment of density between the various albumin concentrations was decreased in order to refine the resolution of the gradient. In this way the morphology of the HSC in the mouse could be identified since fractions can be obtained in which 1 out of 3 cells appears to be an HSC (according to the spleen colony assay).

Table 4

30-Day survival percentage of lethally irradiated rhesus monkeys followin take of bone marrow Fraction 2, rich in allogeneic stem cells

	30-Day survival percentage	Mean survival time in days
Unfractionated marrow 4×10^8 cells/kg*	0%	13 (7 - 25)
Fraction 2 $0.4\text{-}0.5 \times 10^8$ cells/kg**	80%	36 (21- 56)

* 25 animals ** 10 animals

Table 5. Value of the different separation techniques

	Analytical	Preparative	Advantage	Disadvantage
VELOCITY sedimentation	++	+	technique is not compli- cated	sedimentation time-consuming; very large chambers for large cell numbers
CONTINUOUS gradient	++	?	small incre- ments of density can be investigated; separation is fast	collection of fraction and estimation of their density laborious
DISCONTINUOUS gradient	+	++	high separation efficiency; preparation of gradient and collection of fractions relatively simple	recovery of haemopoietic stem cells moderate

Identification could never have been made by investigation of unfractionated bone marrow suspensions, since the stem cell frequency in that population is too low (1 out of 250).

CONCLUSIONS

The conclusions about the value of the analytical and preparative possibilities of the separation techniques discussed above are summarized in Table 5. The velocity sedimentation technique and the continuous albumin gradient designed by Shortman appear to be excellent analytical methods, whereas the discontinuous gradient is preferentially used for preparative purposes. As was already mentioned, the fractionation techniques described above are based on differences in volume or density of the cells being separated. It is likely that those physical properties of the cells differ from species to species, so that adaption of the separation methods is necessary at least in the case of preparative experiments. A disadvantage of the velocity sedimentation technique might be the relatively long separation time which may affect the viability of the different cell populations. Separation by the continuous density gradient technique is quite fast, which is also the case when the discontinuous gradient is used. A real advantage of the latter technique is its simplicity with regard to the preparation of the gradient as well as to the collection of the cell fractions.

References

1. Bekkum, D.W. van and de Vries, M.J. *Radiation Chimaeras*, Logos/Academic Press, London (1967), Chapter II, p.20.
2. Bekkum, D.W. van, *Transplantation 2* (1964) 393.
3. Dicke, K.A., *Bone marrow transplantation after separation by discontinuous albumin density gradient centrifugation*. Thesis, Leiden (1970).
4. Miller, R.G. and Phillips, R.A., *J. Cell Physiol. 73* (1969) 191.
5. Amato, D., Hergsagel, D.E., Miller, R.G., Phillips, R.A.*et al.* *Review of bone marrow transplants at the Ontario Cancer Institute. Transplantation Proc. 3* (1971) 397.
6. Phillips, R.A., Gorczynski, R.M. and Miller, R.G., *Proceedings of a Workshop on Separation of haemopoietic cell suspensions,* held in Rijswijk, 1970 (eds., D.W. van Bekkum and K.A. Dicke) (1971) p. 89.
7. Till, J.E. and McCulloch, E.A., *Radiation Res. 14* (1961) 213.
8. McCulloch, E.A., *Rev. Franc. Etud. Clin. Biol. 8* (1963) 15.
9. Ferrebee, J.W. and Geiman, Q.M., *J. Infect. Dis. 78* (1946) 173.
10. Leif, R.C. and Vinograd, J., *Proc. Nat. Acad. Sci. (Wash.) 51* (1964) 520.
11. Shortman, K., *Austral. J. Exp. Biol. Med. Sci. 46* (1968) 375.
12. Shortman, K., as for ref. 6, p. 43.
13. Shortman, K., *J. Cell Biol. 42* (1969) 783.

14. Dicke, K.A., van Hooft, J.I.M. and van Bekkum, D.W., *Transplantation 6* (1968) 562.

15. Dicke, K.A., Tridente, G. and van Bekkum, D.W., *Transplantation 8* (1969) 422.

16. Dicke, K.A. and van Bekkum, D.W., *Exp. Haemotol. 20* (1970) 126.

17. Dicke, K.A., as for ref. 6, p. 167.

18. Dicke, K.A. and van Bekkum, D.W., *Transplantation Proc., 3* (1971) 666.

19. Dicke, K.A., *Rev. Europ. Etudes Clin. Biol. 15* (1970) 305.

20. Simonsen, M. and Jensen, E., *The graft versus host assay in transplantation chimaeras.* In *Biological problems of grafting* (ed. F. Albert and G. Lejeune-Ledant), Blackwell, Oxford (1959), p.214.

21. Pluznik, D.H. and Sachs, L., *J. Cell. Comp. Physiol., 66* (1965) 319.

22. Bradley, T.R. and Metcalf, D., *Austral. J. Expt. Biol. Med. Sci. 44* (1966) 287.

23. Dicke, K.A., Platenburg, M.G.C., and van Bekkum, D.W., *Cell Tissue Kinet. 4* (1971) 463.

24. Bekkum, D.W. van and Dicke, K.A. (eds.) *Proceedings of a Workshop/Symposium on* in vitro *culture of haemopoietic cells,* held in Rijswijk, 1971 (1972).

20 CELL SEPARATION BY SIZE*

A.M. Denman and B.K. Pelton
Clinical Research Centre,
Watford Road,
Harrow,
Middlesex HA1 3UJ, U.K.

Lymphoid cells may be separated at 1 g in sedimentation chambers. The cell layer is introduced into the chamber on a shallow gradient of Ficoll which stabilizes the initial layer and the fractions which form subsequently. Separation proceeds for 1-18 h depending on the design of the experiment. Lymphocyte populations with distinctive physical kinetic and immunological properties are separated primarily on the basis of cell size although cell density makes a minor contribution. Separation is particularly effective and rapid when 'rosette' formation of sub-populations is achieved before fractionation. Cell recovery and function are very satisfactory.

There is considerable interest in simple physical methods of separating cell mixtures into discrete populations. Such preparative steps have obvious advantages in a variety of biological experiments. Fractionation can be achieved by utilizing differences in density between cells (1,2). These techniques are rapid and simple but the results are not always reproducible because separation is greatly influenced by the physical properties of the materials used for making the necessary gradients. Small changes in pH or salt concentration have a considerable influence on the pattern of cell separation which can be achieved. Different cell types differ in volume as well as in buoyant density, and indeed differences in the former physical property are more striking and should be easier to exploit in separation techniques. Cell separation based on differences in volume has received a great impetus from the work of Miller and Phillips (3,4), and their techniques of velocity sedimentation have been used to achieve a remarkably refined degree of separation of cells with varying biological and, particularly immunological, functions. This paper describes our adaption of Miller and Phillips' methods and illustrates its applicability to a variety of problems of medical and immunological importance.

* *The topic of Ficoll density gradient centrifugation as developed by Drs. Asherson and Zambola for separating lymphoid cells was also covered in the Symposium paper (G.L. Asherson and A.M. Denman). - A series of step gradients are formed in a centrifuge tube and cells placed near the bottom. Following centrifugation the cells accummulate at the interfaces. When paraffin oil or thioglycollate-induced peritoneal exudates are used in the mouse this method provides 'clean' macrophages but the lymphocyte-enriched layer is contaminated with macrophages, polymorphs and red cells.*

Cells fall through fluid under the influence of gravity at a velocity determined by:-

$$s = \frac{2}{9} \frac{(\rho - \rho^1)}{\eta} gr^2$$

where s is the sedimentation velocity, η the coefficient of viscosity, ρ and ρ^1 the densities of the cell and fluid medium respectively, *g* the acceleration due to gravity and r the radius of the cell. In principle the cells to be separated are introduced into the bottom of a sedimentation chamber, followed by a shallow, continuous density gradient which raises the cells as a sharply defined, thin band to the top of the chamber. The cells sediment under gravity at rates determined by their volume. The gradient has two purposes, firstly to allow the cells to form a uniform band during loading and secondly to reduce disruption between cell layers from convection effects once the cells have separated.

APPARATUS AND METHODS

Apparatus

We have used glass apparatus[†] because this is easier to clean and sterilize than other materials. This is particularly important for experiments in which the separated cells are needed for *in vitro* tissue culture experiments. The tubing is silicone rubber which is non-toxic for lymphoid cells and to which such cells do not on the whole adhere.

The main part of the cylindrical sedimentation chamber used in most experiments (Fig. 1, H) is 14 cm in diameter and 10 cm deep. This tapers at the bottom at an angle of 30° to the inlet (Fig. 1, I). The chamber is flanged at the top (Fig. 1, J) and on this is placed a flanged glass lid tapering at a similar angle towards the outlet (Fig. 1, K). Silicone vacuum grease provides an adequate, water-proof seal. It is essential to prevent turbulence at the inlet. For this purpose a porcelain disc, 3 cm in diamteer, and perforated with 2 mm holes, is placed over the inlet and and held in place by a stainless steel ring of similar diameter, 5 mm thick, and 5 mm high. The space within the ring is filled with a single layer of glass beads, 4 mm in diameter. The inlet is connected by a three-way tap (Fig. 1, E) to one tube leading to the chamber for introducing the sample (Fig. 1, F) and by another tube to the gradient maker (Fig. 1, F_1). Both tubes are provided with flow-rate regulators (Fig. 1, D). The tubing is 8 mm in diameter with the exception of F_1 which is 3 mm in diameter. The gradient in all experiments has been based on the buffered step gradient of Miller and Phillips (3). The diameters of the bottles in the gradient maker are such that the levels in the chambers are the same when A and A_1 (Fig. 1) are filled with 500 ml and B (Fig. 1) with 50 ml of gradient material. C in Fig. 1 represent magnetic stirrers.

†*Our original apparatus was constructed by Rudolph Grave AB, Fack 171 20 Solna 1, Stockholm, Sweden.*

Method

The connecting tubing is primed via B with buffered salt solution, most commonly Earle's or Hanks solution, and an additional 30 ml is allowed to enter the sedimentation chamber. This overlay of buffer above the cell band prevents surface effects, reduces the number of cells sticking to the

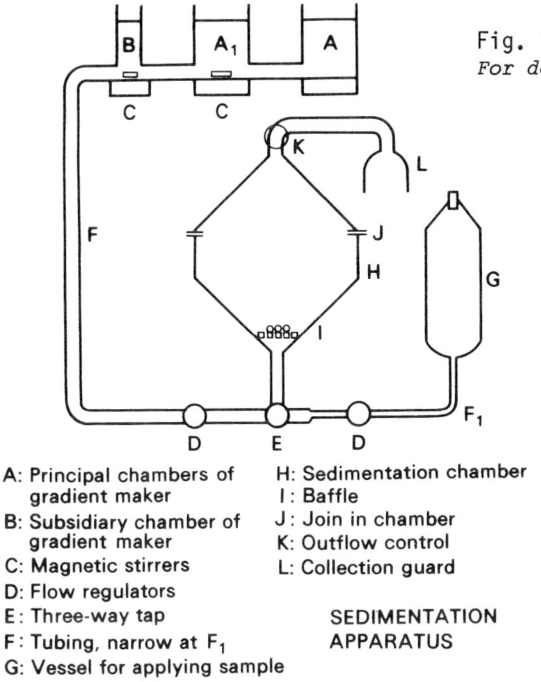

Fig. 1. The apparatus.
For description, see text.

inlet during loading and gives a sharp upper edge to the cell band. After priming with buffer the gradient maker vessels are filled with gradient material but the connections between these vessels are left closed. For most purposes we use Ficoll in Earle's salts medium (Flow Laboratories, Irvine, Scotland), reinforced with 4 per cent heat-inactivated foetal calf serum. This concentration of serum has a negligible effect on density but is valuable in improving cell viability. The concentration of Ficoll is 2.0 per cent in chamber A, 1.0 per cent in chamber A_1, and 0.33 per cent in chamber B. The resulting 'buffered step' gradient is illustrated in Fig. 2.

The cell sample is introduced into G (Fig. 1). If the sample is introduced via B, cells remaining in the tubing become admixed with the gradient material and small numbers become diffused throughout the gradient beneath the main cell band. For this reason it is preferable to load the cells via a separate chamber. The cells are introduced in a 20 ml volume of 0.16 % Ficoll. Most of our work has been concerned with lymphoid cell suspensions and has confirmed the experience of Miller and Phillips (3) with regard to the number of such cells it is possible to separate. The maximum cell number which can be introduced into a sedimentation chamber of the dimensions described is 300 x 10^6 in a concentration not exceeding 15 x 10^6 cells per ml, if streaming is to be avoided.

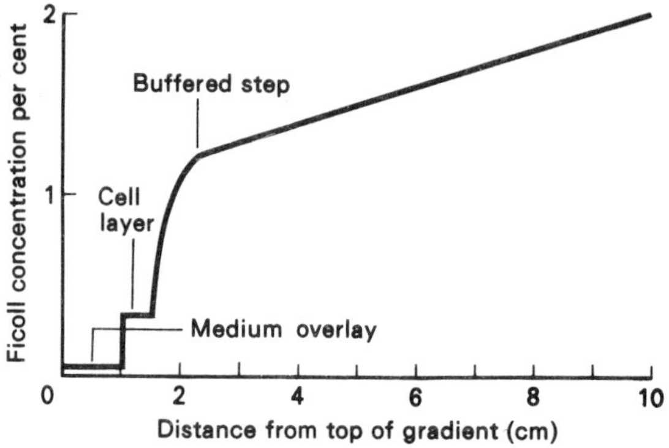

Fig. 2. Buffered step Ficoll gradient for velocity sedimentations. *For 'buffered step' principle, see ref. 3.*

The cells are allowed to run into the sedimentation chamber by gravity as far as they will go, so that the liquid in the upper part of the connecting tube, F_1, is now replaced by air. Vessel G is refilled with 20-30 ml of 0.16 per cent Ficoll in the medium, and the new head of fluid drives the remaining portion of the cell suspension into the chamber. The tubing is thin enough to allow a bubble of air to remain between the tail of the cell sample and the new medium, so that tap E can be closed at the precise moment when the last of the cell sample enters the chamber. At the same time, this manoeuvre avoids diluting the cell sample with more medium. Another precise way of loading the sample which minimizes loss of cells is to close D, and inject the 20 ml sample through a needle into the tubing where it connects with E. The sample can then be washed into the chamber with 5-10 ml of medium from vessel G.

At this stage, the gradient vessels are connected and the gradient formation is started. The first 30 ml, that is the capacity of the connecting tubing, is directed via F_1 into vessel G so that the residual buffer is displaced. The gradient is then directed into the sedimentation chamber at a rate not exceeding 5 ml per min until the cell band has been lifted clear of the conical intake of the chamber and has risen at least 2 cm into its cylindrical portion. Thereafter the flow rate can be safely increased to 20 ml per min. Greater speeds of filling, particularly in initial stages, lead to jet effects which disrupt both the gradient and the cell layer. It is sufficiently accurate and convenient to fill the chamber by gravity, if the gradient makers are 25 cm or more above the intake. Equally good results are achieved if the gradient is pumped into the chamber at a constant rate which allows one to leave the process unattended. Peristaltic pumps are unsuitable for this purpose since the to-and-fro motion of the fluid entering the chamber disrupts the gradient.

The gradient is allowed to enter the chamber until the upper fluid level is 1 cm below the connection between the main chamber and the lid. A cushion of 30% sucrose solution in buffered salt solution is run in from B, filling the intake cone of the chamber and thereby bringing the overlying gradient level with the top of the main separation chamber.

The length of time for which the cells are allowed to sediment is, of course, dependent on the nature of the cell sample. The chamber is emptied by upward displacement with 30% sucrose, which is added via the gradient maker. It is convenient to colour this solution with a few drops of 1% methylene blue so that the point at which the gradient is emptied can be accurately noted. Successive 20 ml fractions are collected from a hooded outlet which protects the samples from bacterial contamination. The intermediate tubing is kept as short as possible to reduce loss of cells. The chamber can safely be emptied at the rate of 20.0 ml per h. This method of emptying is recommended when there are comparatively small volume differences between the cell populations which are being separated. It is more convenient to empty the chamber by gravity, but ideally this alternative procedure should be reserved for experiments in which there are gross differences in cell size within the sample, as admixture between layers is greater.

Technical Variations

Cell numbers

The maximum number of cells which can be separated by this method is governed by the need to prevent streaming which occurs with cell concentrations already noted. Increasing the volume of the cell sample gives a deeper starting band and a poorer separation, so that the absolute number of cells which can be applied is similarly limited. However, greater numbers of cells can be applied if the chamber is widened proportionately. On the other hand, it is undesirable to load a chamber of the dimensions described with cells in a concentration below 2.5×10^6 per ml or, in other words, a total sample below 50.0×10^6 cells. For this there are two reasons: firstly cell recovery is disproportionately impaired with low starting numbers, and secondly viability is less satisfactory. These points are considered later. Relatively small samples of cells are available from human subjects, whether healthy or diseased. It is necessary therefore to separate such specimens without facing the risks of poor recovery and viability of cells in the various fractions. For this purpose the chambers can be scaled down to a diameter of 5 cm without prejudicing the efficacy of the procedure. The cell load must be reduced proportionately, but viability and recovery are improved because the cell concentration per fraction is comparable with that obtained in larger chambers. In this way 10.0-40.0 x 10^6 cells can readily be separated.

Duration of experiment

Separations can be carried out over a more prolonged period in longer chambers. We have used cylinders up to 20 cm high for overnight separation This permits more flexible experimental protocols and can sometimes improve the results. Satisfactory fractionations have been attained after sedimenting the cells for 18 h.

Gradient material

Ficoll has proved the most satisfactory material for gradients but we, in common with other workers, have used crude bovine serum albumin (Fraction V) or foetal calf serum with equally good results. The choice of fluid is determined by the nature of the experiment. For short-term separations, not exceeding 4 h in duration, buffered salt solution is adequate. For longer separations or experiments involving *in vitro* culture of lymphoid cell fractions the cells survive better in tissue-culture medium containing a low concentration of foetal calf serum, namely 3-4%. Solutions containing Ca^{2+} ions cause some clumping of lymphoid cells but this is of little consequence unless cells are deliberately aggregated before loading in order to isolate cells with specific biological characteristics (see below). In this type of experiment, non-specific clumping can be reduced if the cells are prepared and sedimented in the following buffer:-

Stock solution:- NaCl, 76.0 g; KCl, 2.0 g; NaH_2PO_4, 0.5 g; Na acetate, 15.0 g; K_2HPO_4 1.0 g; glucose, 10.0 g; pH 7.2.

Make up to 1000 ml with distilled water, and *dilute* x 10 with distilled water before use.

Viability is not affected, provided the period of sedimentation does not exceed 4 h which is anyway time enough for cell aggregates to separate.

Clearly the density of the gradient material cannot be reduced below the figures cited; but it is sometimes advantageous to increase proportionately the concentration of the material in each vessel of the gradient maker. This modification enables chambers of standard depth to be used for longer runs; in some experiments whose object is to isolate a population of cells uniformly much larger than the remainder in the sample, this fraction is less contaminated with smaller cells if the density of the gradient is increased, whilst its other features are retained. We have used 2-10% Ficoll gradients for this purpose. Of course, cells are no longer separated entirely on the basis of cell volume; but this is of no practical importance. Indeed there is evidence that both these physical parameters are involved in separation by velocity sedimentation (5).

Factors affecting recovery

The factors which affect cell recovery are mechanical and biological. Losses from the separation procedure can be minimized by using short lengths of silicone rubber tubing between various glass chambers, and by concentrating the cells from each fraction in plastic containers with conical bottoms (Sterilin Ltd., Richmond, Surrey) using a large, stable centrifuge such as the Mistral 6L. These measures reduce the risk of dispersing small cell buttons recovered from relatively large volumes of fluid during centrifugation. There is no difficulty in spinning down the cells in the low-density media employed in sedimentation techniques. Recovery averages 70% in experiments with cells whose total number exceeds 50 x 10^6. This figure is reduced to around 20% if cells numbering below 20 x 10^6 are separated in the larger chambers, but is not restored to acceptable values if one uses scaled-down chambers. Recovery is not improved by siliconizing the glassware. Rapidly proliferating populations or biologically fragile cells are sometimes more vulnerable than other cell types during prolonged separations, so that in some experiments recovery may be reduced after overnight runs.

Factors affecting viability

Cell viability is highly satisfactory after velocity sedimentation. Not only has this been demonstrated by *in vivo* transfer of lymphoid cells (4, 6,7) but also by *in vitro* studies of these cells from humans, rats and mice. The cells transform in response to mitogenic or antigenic stimuli, incorporate RNA and DNA precursor isotopes, produce antibody measurable by plaque techniques, are cytotoxic for appropriate target cells, and in experiments involving antibody synthesis *in vitro* (8,9) perform normally in this stringent test of functional integrity (10).

Viability as well as recovery is impaired in fractions whose concentration falls below certain limits, and is usually the consequence of applying samples containing less than 2.0 x 10^6 cells per ml. Recovery in terms of function and cell numbers can be improved by adding 'carrier'cells whose volume is roughly equivalent to that of the population which one wants to isolate; cell concentration in the relevant fractions is thereby improved. For example, the sedimentation velocity of sheep erythrocytes overlaps with that of mouse small lymphocytes, whilst large cells from human lymph nodes co-sediment with many fibroblast cell lines. The importance of cell density is evident in many *in vitro* biological systems, so that its applicability in this context is hardly surprising.

Temperature

We run the majority of our separations at 4° because cell viability is optimal at this temperature, particularly if the procedure is prolonged for more than 6 h. Nevertheless if the same cell sample is run at 4° and at room temperature (approximately 18°) separation proceeds more efficiently and more rapidly at the latter temperature; Coulter analysis of volume distribution reveals that lymphoid cells shrink at low temperatures. Thus, at 4° there is a 10-20% retardation in sedimentation. Moreover at room temperature, unlike 4°, the formation of discrete bands is clearly visible, indicating an improved resolution. Indeed when blood mononuclear cells which have been separated by density gradient centrifugation in a Ficoll-isopaque mixture (11) are sedimented at room temperature, three obvious bands are formed: the upper, narrow band consists of residual Ficoll dissociated from the mononuclear cells within the chamber, the middle one of red cells, whilst the lower more diffuse band contains the lymphoid cells.

ILLUSTRATIVE SEPARATIONS

In general lymphoid cells sediment according to a predictable pattern. That found for mouse spleen cells is depicted in Table 1.

Table 1. Sedimentation Velocity of Mouse Spleen Cells

Cell type	Velocity *mm per h*
red cells	2
small lymphocytes	2 - 4
medium lymphocytes	4 - 6
large lymphocytes, macrophages	5 - 8
'rosettes' - single attached layer	6 - 9
'rosettes' - 'mulberry' type	8 - 14

Velocity sedimentation has numerous applications in biology including cell synthronization (12) and the separation of haemopoietic stem cells from bone marrow (6). This method has been applied extensively to the separation of mouse lymphoid cells with varying immunological functions using *in vivo* adoptive transfer systems for their detection. Illustrative examples of these and other applications are now given.

Separation of lymphoid cells in DNA synthesis

A representative separation is shown in Fig. 3, for cells from normal thymus.

Separation of immunologically active populations of lymphoid cells

Lymphocytes with surface immunoglobulin receptors for antigens such as sheep red cells can be detected by allowing the red cells to bind to these antigen-sensitive cells. These lymphocytes are surrounded by a corona of red cells which can be recognised microscopically as a cell 'rosette' (13). These rosettes can readily be separated by velocity sedimentation.

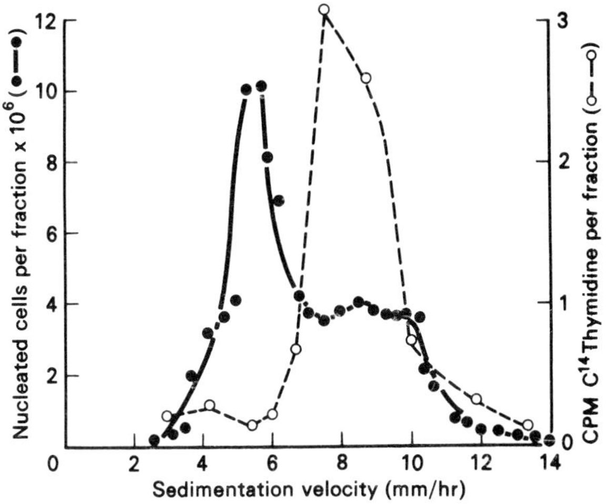

Fig. 3. Sedimentation profile of DNA synthetic cells in normal mouse thymus. *200.0 x 10^6 A/Snell mouse thymus cells, sedimented for 6 h at 4°, 0.33-2.0% Ficoll gradient. 20 ml fractions labelled with ^{14}C-thymidine, 1.0μC per 2.0 x 10^6 cells* in vitro *for 4 h.*

Lymphocytes which secrete antibody to sheep red cells bind multiple layers of these cells and the resulting aggregates sediment still more rapidly. After separation the specificity of the reaction can be confirmed by demonstrating that the fractions containing antigen-binding cells are enriched for antibody-forming cells detectable by the well-known Jerne plaque technique. An example of this application is shown in Table 2.

Table 2. Immunological responses to sheep red blood cells by subpopulations of mouse spleen cells

RFC = rosette-forming cells, PFC = plaque-forming cells

A/Snell mice were immunized with 2.0 x 10^8 sheep red cells. Eight days later 2.0 x 10^8 spleen cells mixed 10^9 red cells were sedimented for 4 h in a 0.33-2.0% Ficoll gradient in Ca^{2+} free buffer. Fractions were examined for their content of preformed RFC with single or multiple attached layers of indicator cells, and were assayed in vitro *for PFC-producing 19 S or 7 S antibody to sheep red cells. RFC in Pool 6 were predominantly bound to multiple layers of sheep red cells.*

SEDIMENTATION VELOCITY *mm per h*	POOL NO.	% CELLS RECOVERED	% RFC	PFC per 10^6 CELLS 19S	7S
2.0-3.9	1	18.5	0	0	31
3.9-4.5	2	24.6	0	0	60
4.6-6.0	3	15.2	0	19	75
6.1-8.0	4	16.3	0	12	119
8.1-9.6	5	20.4	37	101	410
9.7-16.0	6	5.0	80	30	540

Isolation of virus-infected cells

The study of virus infections may sometimes be facilitated by techniques which allow the separation of infected from non-infected cells. Cells with membranes to which haem-adsorbing virus has adsorbed or from which it is budding form rosettes with suitable indicator red cells; the resulting cell aggregates can be readily separated by velocity sedimentation. An example is given in Fig. 4, taken from data obtained in a current study of virus-lymphocyte interaction.

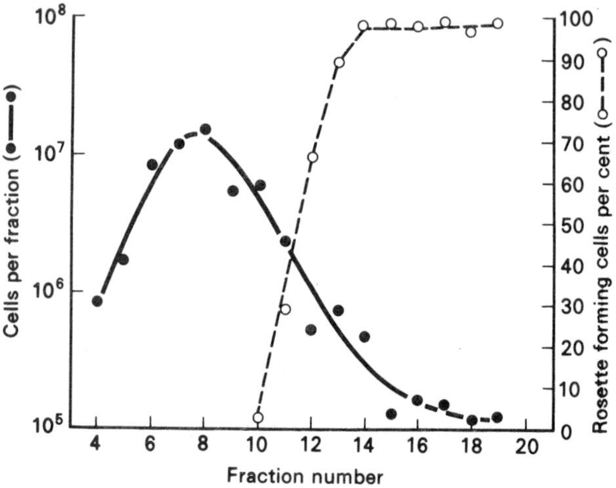

Fig. 4. Separation of lymphoid cells infected with haem-adsorbing virus by rosette formation with chicken RBS. *CBA mouse spleen cells were exposed to Sendai virus, 10^5 HA units per 2.0 x 10^6 cells at 37° for 1 h. 2.0 x 10^8 spleen cells mixed mixed with 1.0 x 10^9 chicken red cells were sedimented for 4 h in a 0.33-2.0% Ficoll gradient in Eagle's medium.*

Characterizing normal and abnormal populations of peripheral blood mononuclear cells

There are a variety of situations in which it is clinically important to analyse the composition of the mononuclear cells circulating in the blood. The recent demonstration of heterogeneity among these cells (14) gives added interest to such analyses. For example, patients with immune deficiency diseases can be expected to lack certain lymphocyte populations, whereas in chronic inflammatory diseases abnormal populations appear in the blood (15). The mixed nature of the blood mononuclear cells in normal peripheral blood is illustrated in Fig. 5 and Table 3.

Fractions were tested for their response to the mitogen phyto-haemagglutinin and their ability to kill target cells coated with antibody (16). Such tests are of value in distinguishing lymphoid cell populations with differing life-span and function. The sedimentation pattern in Fig. 6 emphasizes that it may be necessary to increase the period of separation in order to reveal sub-populations of lymphoid cells with different functional properties (Table 3).

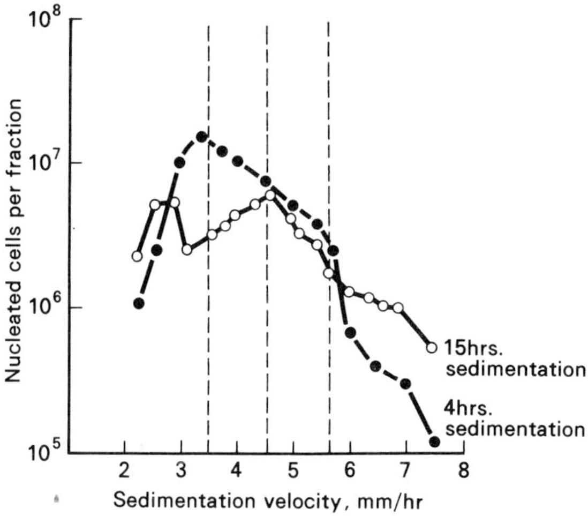

Fig. 5. Sub-populations in human peripheral blood mononuclear cells. *Mononuclear cells were obtained from heparinized samples of normal human blood by 'plasmagel' separation (Roger Bellon, Neuilly, Seine, France) followed by centrifugation on a Ficoll-isopaque density gradient* (11). *80 x 10^6 Cells were sedimented for 4 or 15 h on a 0.33-2.0% Ficoll in Eagle's medium gradient at 4°. Fractions were analysed by Coulter counter, and adjacent fractions were pooled and tested for transforming ability with phytohaemagglutinin, and for cytotoxic action on chicken erythrocytes sensitized with rabbit anti-chicken red cell antibody.*

The success of separation procedures can be monitored by determining the volume distribution of the cells in each fraction by Coulter counter analysis. The different picture in the slowly and rapidly sedimenting fractions from the experiment of Fig. 5 is illustrated in Fig. 6.

Isolation of abnormal cells from blood and other biological fluids

We have found velocity sedimentation a useful technique for isolating abnormal cells from such sources as joint effusions from patients with chronic inflammatory arthritis and abnormal peripheral blood. This is an essential preparative step in a variety of experiments in which the properties of such cells are studied. An example of this application of the technique is provided in Fig. 7; the atypical mononuclear cells in the blood of a patient with infectious mononucleosis (glandular fever) were successfully isolated, the lower fractions containing atypical mononuclear cells exclusively which were morphologically and biochemically characteristic of the disease.

DISCUSSION

Velocity sedimentation is applicable to a variety of immunological and biological problems necessitating the isolation of cell populations. The advantages and disadvantages can be summarized.

[continued on p. 197

Table 3. Sub-populations in human peripheral blood mononuclear cells.

Vertical broken lines in Fig. 5 correspond to the pooled cell fractions assayed in Table 3.

Transformation test: *1.0 x 10^6 cells cultured for 72 h in 20% RPMI 1640 80% Eagle's medium with 10% foetal calf serum. ^{14}C-Thymidine, 0.08 μC, added 24 h before harvesting.*

Cytotoxicity test: *5.0 x 10^5 lymphoid cells exposed to 1.0 x 10^5 ^{51}Cr-labelled chicken erythrocytes sensitized with rabbit antibody to chicken red cells. Counts are ^{51}Cr release corrected for background loss after 18 hours'incubation (16).*

SEPARATION TIME	SEDIMENTATION VELOCITY mm per h	CELL NUMBERS PER POOL x 10^6	PHA TRANSFORMATION cpm per 10^6 cells cultured	CYTOTOXICITY per 5.0 x 10^5 cells
4	2.0 - 3.5	9.4	4005	6.6
	3.6 - 4.7	30.8	7415	13.1
	4.8 - 9.4	12.7	3728	23.7
15	2.2 - 3.5	9.9	1564	4.8
	3.6 - 4.7	10.8	19752	20.5
	4.8-- 5.7	12.1	3323	12.6
	5.8 - 7.0	15.2	1564	41.5

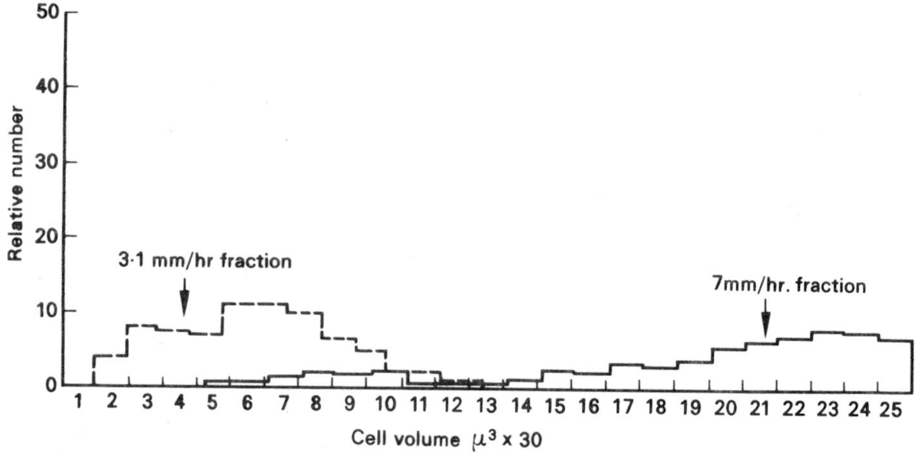

Fig. 6. Cell volume distribution (Coulter counts) in representative fractions of peripheral blood mononuclear cells. *The volume distribution is shown for slowly (3.1 mm/h) and rapidly (7 mm/h) sedimenting cells.*

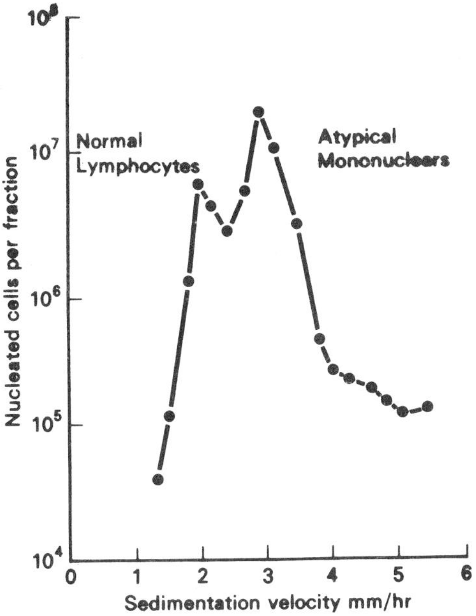

Fig. 7. Separation of atypical mononuclear cells from the blood of a glandular fever patient. *70.0 x 10^6 mononuclear cells isolated from blood of patient V.P. aged 18 yrs. with typical infectious mononucleosis by the technique described in legend to Fig. 5, and sedimented for 14 h on 2-10% Ficoll in Eagle's medium gradient.*

Advantages

The procedure is simple to learn and operate. The apparatus is easily constructed and indeed one commercial source is now available (Johns Glass Company, Toronto). The results are completely reproducible because the usual source of variation in separation experiments, namely the material employed in the density gradient, plays a minor role in velocity separation. The method is remarkably efficient in separating cells with minor differences in morphology, a point which is particularly relevant to the study of lymphoid cell populations in man and experimental animals. With the precautions mentioned, recovery and viability are both excellent.

Disadvantages

Compared with density separation, this method is prolonged, and vulnerable cells are exposed to lengthy *in vitro* manipulations. The separation of markedly heterogeneous populations is poorer than that obtained with more homogenous samples. The commonest problem with such samples is to find that whilst smaller cell fractions are fairly homogeneous, the more rapidly sedimenting fractions are contaminated with small cells. This difficulty can be mitigated in two ways. The first is to carry out a preliminary fractionation in a continuous or discontinuous density gradient. The second is to sediment the sample in one chamber for a short period and to re-sediment the important fractions in a second chamber. Granulocytes in large numbers complicate separations in another way, since they tend to clump with cells of the same or different type despite the liberal use of heparin; a preliminary density gradient step is essential.

The problem of cell numbers is of little importance, since chambers of greatly increased diameter can be utilized where necessary. Contaminating red cells, which would otherwise reduce the number of lymphoid cells which can be loaded without streaming, are easily lysed with NH_4Cl-Tris buffer (17).

Although rosettes induced by viral haemadsorption remain stable throughout the 4 h separation procedure, immune rosettes are not entirely stable. This instability can be demonstrated by mixing the small lymphocytes in the upper fractions with further red cells from the same source, whereupon more rosettes are formed, usually totalling 10-20% of the number produced on initial mixing. This is confirmed by experiments on tolerance formation. Lymphocytes can be made unresponsive to red cell antigens by forming rosettes with the appropriate red cells and then removing those rosettes by density gradient centrifugation; the remaining cells are thereby depleted of specifically reactive lymphocytes (18). Comparable results have been claimed using velocity sedimentation, but in similar experiments (19) the rosette-depleted lymphocytes consistently gave an antibody response of about 20-30% of control values, judged by counts of plaque-forming cells.

Acknowledgements

The authors are grateful to Dr. G. Möller, the Wallenberglaboratory, Stockholm, and Dr. Simon Hunt, Sir William Dunn School of Pathology, Oxford, for helpful discussions.

References

1. Raidt, D.J., Mishell, R.I. and Dutton, R.W., *J. Exp. Med. 128* (1968) 681.
2. Shortman, K., *Aust. J. Exp. Biol. Med. Sci. 46* (1968) 375.
3. Miller, R.G. and Phillips, R.A., *J. Cell Physiol. 73* (1969) 191.
4. Phillips, R.A. and Miller, R.G., *Cell Tissue Kinet. 3* (1970) 263.
5. Haskill, J.S. and Marbrook, J., *J. Immunol. Methods 1* (1971) 43.
6. Worton, R.G., McCulloch, E.A. and Till, J.E., *J. Cell Physiol. 74* (1969) 171.
7. Miller, R.G. and Phillips, R.A., *Proc. Soc. Exp. Biol. Med. 135* (1970) 63.
8. Osoba, D., *J. Exp. Med. 132* (1970) 368.

9. Sjoberg, O. and Denman, A.M., unpublished observations.

10. Mishell, R.I. and Dutton, R.W., *J. Exp. Med. 126* (1967) 423.

11. Böyum, A., *Scand. J. Clin. Lab. Invest. 21, suppl. 97* (1968) 1.

12. Macdonald, H.R. and Miller, R.G., *Biophys. J. 10* (1970) 834.

13. Zaalberg, O.B., *Nature (Lond.) 202* (1964) 123.

14. Wilson, J.D., and Nossal, G.J.V., *Lancet, ii* (1971) 788.

15. Horwitz, D.A., Stastny, P. and Ziff, M., *J. Lab. Clin. Med. 76* (1970) 391.

16. Perlmann, P., and Holm, G., *Adv. Immunol. 11* (1969) 117.

17. Boyle, W., *Transplantation 6* (1968) 761.

18. Bach, J.F., Muller, J.Y. and Dardenne, M., *Nature (Lond.) 227* (1970) 1251.

19. Denman, A.M. and Sjoberg, O., unpublished observations.

21 COMMENTS and SUPPLEMENTARY CONTRIBUTIONS

Editor's Note - *This Section comprises comments on topics dealt with earlier in the book, and brief reinforcing contributions. The material in this Section, as in the preceding Sections, arises largely from the British Biophysical Society's Symposium (December 1971) but should not be regarded as a comprehensive record of the Symposium; it includes some techniques which pertain to analytical rather than preparative biochemistry.*

ELECTROPHORETIC METHODS *(cf. Articles 1-9)*

Temperature gradients in electrophoresis

J.O.N. Hinckley *(Transphoresis Ltd., Brandenburg Ltd., U.K.)*

Concerning distortion of electrophoretic zones due to temperature gradients normal to the electrophoretic current, it has been suggested that the temperature gradient may be calculated quite simply. A simple calculation may be a good approximation only at low ratios of power to cooled area. For higher ratios, however, the higher temperature gradient will reduce the resistance of central electrolyte more than peripheral electrolyte, as electrolytes have a negative temperature coefficient of resistance. This in turn will increase the central current density at the expense of the peripheral current (although overall current is maintained constant) - a process which itself leads to a further increase in temperature gradient and current density gradient. Calculation of these gradients in the steady state, as now computer-programmed in our laboratory, involves Bessel functions, a conclusion reached also in another laboratory (1). Resultant zone distortion would therefore be worse than simple theory predicts and the gradients would cease to be strictly parabolic. (Similar considerations apply to transphoresis (displacement electrophoresis, isotachophoresis; see below), in calculating the steady-state non-degenerating profile of the inter-species boundary or 'front'.) An important feature of our calculations is that for any given steady-state power dissipation per unit length of column, where the column wall is ideally thermostatted and column-end effects are disregarded, the temperature gradient discussed is independent of area for cross-sections of similar geometry, at least for first-order effects. Thus, for a given power dissipation per unit length, and a given central temperature increment, the area of cross-section will be roughly in inverse proportion to total ion constituent concentration.

Editor's Note - Mr. Hinckley's comments take account of remarks made at the Symposium by C.J.O.R. Morris and A.R. Williamson. A paper on temperature measurement and control in gels has recently appeared (2), and Article 2 (A. Brownstone) earlier in this book is also relevant. In answer to a question (H. Hillman) on the effect of ohmic heating on recovery of proteins from gels, the importance of efficient cooling was stressed (C.J.O.R. Morris; S. Hjertén), and ribbed cooling plates were suggested (H.S. Bingley).

1. Jakob, M., *Trans. Amer. Soc. Mech. Engrs. 70* (1948) 25.

2. Blattler, D.P. & Bradley, A.R., *Anal. Biochem. 47* (1972) 296.

Toxicity of acrylamide *(cf. Article 2)*

E. Reid and A.R. Jones *(Univ. of Surrey, U.K.)*

The toxicity of acrylamide, to which A. Brownstone draws attention, has been studied by investigators concerned with acrylamide as an industrial hazard through its use in mining operations as a soil-stabilizer. It is a cumulative neurotoxin (1, 2), and a number of cases of acrylamide poisoning in humans have been reported (3, 4), the symptoms of which include ataxia and dermatitis. Consequently persons using acrylamide should wear gloves and take other appropriate precautions (3). The toxicity affects a wide range of experimental animals (5), the LD_{50} for rats being *c.* 200 mg/kg. Smaller doses produce adverse effects such as ataxia, and may be lethal if repeated (2). The toxicity of the monomer may reflect reaction of the ethylene bond with non-protein sulphydryl groups (6). The polymer is stated to be non-toxic (3). N,N'-methylenebisacrylamide ('BIS') is understood not to be a serious hazard.

Whilst no hazard is known to be associated with the low molecular-weight material of the nature of polyacrylamide that appears in the eluates from gel columns *(Articles 2 & 3)*, care should obviously be exercised in work with such eluates.

1. Kuperman, A.A., *J. Pharmacol. Exp. Therap. 123* (1958) 180.

2. Fullerton, P.M. & Barnes, J.M., *J. Indust. Med. 23* (1966) 210.

3. Garland, T.O. & Patterson, M.H.W., *Brit. Med. J. 4* (1967) 134.

4. Auld, R.B. & Bedwell, S.F., *J. Canad. Med. Assn. 96* (1967) 652.

5. McCollister, D.D., Oyen, F. & Rowe, V.K., *Toxicol. Appl. Pharmacol. 6* (1964) 172.

6. Hashimoto, K. & Aldridge, W.N., *Biochem. Pharmacol. 19* (1970) 2591.

Gel techniques in S. Hjertén's laboratory *(cf. Article 4)*

Concerning the availability of the dextran gels (T. Rosenbaum), these have to be prepared in the laboratory by cross-linking dextran with epichlorhydrin, there being no commercial source. *Answers to* P. Marko: the enzymatic digestion is done at room temperature; the use of sodium dodecylsulphate in the dextran gels has been tried; the sucrose gradients were prepared by layering; the separation of subcellular particles has been achieved. Concerning the use of methyl cellulose as an anti-electro-osmotic agent (D.H. Leaback), it may act by increasing the local viscosity near the solid surface and thus diminishing the rate of flow.

Gradient pore electrophoresis

D.H. Leaback *(Inst. of Orthopaedics, Stanmore, Middx., U.K.)*

Several new electrophoretic methods have been developed in recent years which offer more direct indications of certain molecular characteristics of proteins than was possible from earlier techniques. Amongst these, the 'gradient pore' technique (1) offers potential for indicating relative 'size and shape' of proteins in mixtures, and does this as the result of the molecular sieving which takes place when proteins are driven electrophoretically through a gel matrix with a gradient of progressively smaller 'pores' (i.e. of increasing gel density).

Comment by A. Pusztai. - There is a 'gradient pore' technique in which electrophoresis is conducted in phenol-acetic acid aqueous urea solutions across linear gradients of acrylamide (2). In this case as all proteins (or rather subunits in this dissociating solvent) are totally positively charged the mobility is the function of the size and shape of the proteins only.

1. Margolis, J. & Kenrick, K.G., *Anal. Biochem. 25* (1968) 347.

2. Thorum, W. & Mehl, E., *Biochim. Biophys. Acta 160* (1968) 132.

Isoelectric focusing *(cf. Article 7)*

J.S. Fawcett (*replying to* A. Pusztai).- Concerning Ampholine-protein interactions, there is indeed a possibility of complex formation, especially in view of the nature of the Ampholine and the low ionic strength prevailing at the isoelectric point. Isoelectric focusing of proteins thought to be homogeneous has often given multiple zones and has led many researchers to suspect artefacts due to complexing. However, in almost all cases where material from one of these zones has been isolated and re-focused, it has given a single band and not multiple zones as would be expected if complexing had occurred. The notable exception is albumin which, it is claimed, is modified by a minor component in the Ampholine mixture (1). In this case the relative concentration of protein in the zones is dependent on the ratio of protein to Ampholine. In a study of the interaction of L-amino acid oxidase with tritium-labelled Ampholine (*ref. 19 in Article 7*), it was estimated that one mole of enzyme complexed with 1.5 moles of Ampholine. Subsequent disclosures by the manufacturers that acidic and basic amino acids had been added to the Ampholine mixture have thrown doubt on this conclusion, since any L-amino acids present would also be tritiated and, being substrates, would bind to the enzyme.

Concerning possible oxidation of proteins during electrofocusing (*as raised by* A. Brownstone; *cf. ref.* 2 - *Ed.*), there is some evidence that proteins may be oxidized by diffusion of oxygen from the anode electrode. Several workers have added thiol compounds to the system, and at the suggestion of H. Davies the addition of a layer of ascorbic acid solution between the Ampholine and the phosphoric acid at the anode end has been tried out by J.S. Fawcett and colleagues. One must be sure that the proteins themselves are not modified by these chemicals. It is important also to guard against the oxidation of the protein during the detection and isolation procedures that follow the electrofocusing. A loss of enzyme activity will be expected from those enzymes that require the presence of inorganic cations for stability. In some cases these may be re-activated in the presence of cations after electrofocusing.

In the scanning apparatus as described (*Article 7; question by* T. Rosenbaum, *answered by* J.S. Fawcett), is there any effect on the pH gradient as the solution passes the membrane of the electrode vessel? – The density gradient is pumped past the top electrode membrane in the ISCO apparatus, but diffusion of electrode solution through this membrane is very slow and is unlikely to affect the pH gradient. With the modified apparatus as constructed the top electrode dips directly into the tube and no membrane is used.

1. Kaplan, L.J. & Foster, J.F., *Biochemistry 10* (1971) 630.

2. Jacobs, S., in *Protides of the Biological Fluids* (H. Peeters, ed.), Pergamon, Oxford (1971), p. 499.

In the Hannig apparatus (*as mentioned in Articles 7 & 8; remark by* A. Pusztai) the available maximum passage time, $2\frac{1}{2}$ h, could be extended to at least 8-10 h with proper gearing. However (*answer by* J.S. Fawcett), the liquid flow rate would be so slow as to result in very poor stabilisation against convection mixing.

In the electrophoretic purification of viruses (*Article 9; query by* A. Pusztai) there is surely the difficulty that the virus suspension as injected into the flow stream is of higher density and tends to spread, even to the full width of the machine. However (*answer by* L.C. Robinson), such spreading is found only with crude samples, the band being stable and having little spread if purer samples are run.

Direct optical scanning of gel columns *(Editor's note)*.- Progress with the difficult problem of scanning during a gel run has been made by Ackers (e.g. 1).

1. Ackers, G.K., *Adv. Prot. Chem. 24* (1970) 343.

Displacement electrophoresis

Isotachophoresis (Displacement electrophoresis)

D. Peel *(Pye Unicam Ltd., U.K.)*

Isotachophoresis is a new name (1) for a technique which was conceived as long ago as 1919 (2) and has the alternate designation 'displacement electrophoresis'. The technique was first described by Kendall and Crittenden (3). A series of papers by Kendall and his colleagues showed the fundamental properties of the method in the course of separating metal ions (4) and rare earths (5) and suggesting the separation of proteins (6). More recently the method has been applied by several workers to the separation of isotopes (7, 8), inorganic ions (9) and small organic ions and proteins (10 - 15), in various forms of apparatus (16 - 20).

Isotachophoresis is a method of electrophoresis in which a mixture of ions is placed in a tube behind a fast ion (the leading ion) and in front of a slow ion (the terminating ion). The ions in the mixture separate in order of mobility into individual compartments, in direct contact with each other, with boundaries of varying sharpness between neighbours. The ions all have the same sign and preferably have a common counter-ion which for weak ions should be a buffer. A large number of 'discontinuous' buffer systems with appropriate leading and terminating ions have recently been described (21). The leading and terminating ions should be pure since impurities of intermediate mobility move into the sample compartment.

When the separation is complete all ions move at constant speed and the potential gradient in each compartment is inversely proportional to the mobility of that ion. After separation, the concentration of any one ion is constant throughout the compartment and bears a fixed relationship to the concentration of the leading ion irrespective of the starting concentration in the sample (22). The concentration of the leading ion remains constant and can be varied through wide limits. A low concentration allows an increased potential gradient for a given heat output and gives a greater distance between boundaries for a fixed sample size. The length of tube occupied by an ion is directly proportional to the amount of that ion in the sample, so that quantitative results can be obtained by measuring the length of tube occupied by each ion. The final concentration of an ion after separation may be many times greater than the starting concentration. This effect is particularly noticeable with proteins.

Kendall suggested the possibility of adding to the sample, as a marker, a coloured ion of mobility intermediate between that of two ions of interest (6). This concept has been developed by Vestermark (23, 24) and has been used in protein separations (14, 15). The disadvantage of this technique is that it makes the sample more difficult to resolve.

Theoretical studies of isotachophoresis have been published which deal with the pH in each compartment, maximum boundary width, the time required for a given separation and maximum heat output (25, 26). Martin and Everaerts (25, 27) developed a capillary tube apparatus, described elsewhere (19), that permits the use of high potential gradients (100 V/cm), since cooling is efficient and temperature gradients across the tube are small. The sharpness of the boundary is directly proportional to the potential gradient. With a constant current the heat output in each compartment is proportional to the potential gradient and with thin walled tubes the temperature is proportional to the heat output. A fixed thermocouple on the outside of the tube registers the temperature change at boundaries. The temperature record is a staircase shape with the height of a step showing the mobility, and the length the quantity of each ion. This detector responds slowly because of the time taken for the heat to travel through the tube wall to the thermocouple. An optical detector has been shown to give much more detailed patterns than a thermocouple with the same sample (15). However, the optical detector responds to a limited range of components and the record is more difficult to interpret.

Many protein separations by isotachophoresis have been started but not completed by workers using the steady-state-stacking procedure of Ornstein and Davis to concentrate dilute samples in the first stage of acrylamide gel electrophoresis (28, 29). In these conditions the sharpness of separation boundaries is improved by using a gel to restrict diffusion and prevent the density disturbance caused by concentrated protein solutions. Lower potential gradients are used but the apparatus required is simple and the gels may be scanned in a suitable UV densitometer. A gel composition which minimises molecular sieving as recommended for isoelectric focusing should be suitable for isotachophoresis (30). Isotachophoresis is also possible in sieving gels with a terminating ion of low mobility.

1. Haglund, H., *Science Tools 17* (1970) 2.

2. Kendall, J., *personal communication* (1971).

3. Kendall, J. & Crittenden, E.D., *Proc. Nat. Acad. Sci. (Wash.) 9* (1923) 75.

4. Kendall, J. & West, J.A., *J. Amer. Chem. Soc. 48* (1926) 2619.

5. Kendall, J. & Clarke, B.L., *Proc. Nat. Acad. Sci. (Wash.) 11* (1925) 393

6. Kendall, J., *Science (Wash.) 67* (1928) 163.

7. Fiks, V.B., *Russ. J. Phys. Chem.* (English transl.) *38* (1964) 1218.

8. Konstantinov, B.P. & Bakulin, E.A., *Russ. J. Phys. Chem.* (English transl.) *39* (1965) 315.

9. Everaerts, F.M., *Thesis,* Technische Hogeschool, Eindhoven (1968).

10. Everaerts, F.M. & Verheggen, Th.P.E.M., *Science Tools 17* (1970) 17.

11. Everaerts, F.M., *J. Chromatog. 52* (1970) 415.

12. Konstantinov, B.P. & Oshurkova, O.V., *Russ. J. Tech. Phys.* (English transl.) *37* (1967) 1745.

13. Peel, D., Hinckley, J.O.N. & Martin, A.J.P., *Biochem. J. 117* (1970) 69P.

14. Svendsen, P.J. & Rose, C., *Science Tools 17* (1970) 13.

15. Arlinger, L. & Routs, R., *Science Tools 17* (1970) 21.

16. Konstantinov, B.P. & Oshurkova, O.V., *Soviet Phys. Tech. Phys.* (English transl.) *12* (1968) 1280.

17. Preetz, W., *Talanta 14* (1967) 143.

18. Behne, D., Bilal, B.A., Freyer, H.D. & Thiemann, W., *Talanta 15* (1968) 153.

19. Everaerts, F.M. & Verheggen, Th.P.E.M., *J. Chromatog. 53* (1970) 315.

20. Everaerts, F.M., Vacik, J., Verheggen, Th.P.E.M. & Zuska, J., *J. Chromatog. 60* (1971) 397.

21. Jovin, T.M., in preparation. Quoted in ref. 30.

22. Kohlrausch, F., *Annalen Phys. 62* (1897) 209.

23. Vestermark, A., *Naturwiss. 50* (1967) 470.

24. Vestermark, A., *Biochem. J. 104* (1967) 21*P*.

25 Martin, A.J.P. & Everaerts, F.M., *Proc. Roy. Soc. (Lond.) A 316* (1970) 493.

26. Brouwer, G. & Postema, G.A., *J. Electrochem. Soc. 117* (1970) 874.

27. Martin, A.J.P. & Everaerts, F.M., *Anal. Chim. Acta. 38* (1967) 233.

28. Ornstein, L., *Ann. N.Y. Acad. Sci. 121* (1964) 321.

29. Davis, B.J., *Ann. N.Y. Acad. Sci. 121* (1964) 404.

30. Chrambach, A. & Rodbard, D., *Science (Wash.) 172* (1971) 440.

Transphoresis (Displacement electrophoresis)

J.O.N. Hinckley *(Transphoresis Ltd., Brandenburg Ltd., U.K.)*

Concerning Dr. Peel's use of the term *Isotachophoresis* for a technique to which a wide variety of names have been applied, my proposed term *Transphoresis* warrants justification. It is a contraction of 'transfer number electrophoresis', so named because its theory and practice are those of the classic direct or moving boundary method of transfer number determination, on which there are at least 80 papers mostly by MacInnes, Longsworth and co-workers, whence most of the apparatus and detection methods used in transphoresis have been borrowed. Thus the order of analysate region or 'step' alignment, which is the order of decreasing ion constituent transfer numbers, and the uniformity of concentration in these steps, are a consequence of the Kohlrausch regulating function which refers to transfer numbers. Most electrophoresis text-books fail to distinguish this method from the radically different free boundary and zone techniques, which use co-running buffers - a failure I attribute to the lack of emphasis on the strong connection with transfer number determination theory. Nearly all of us as undergraduates did in fact learn and practise the direct method of transfer number measurement, which features in all physical chemistry text-books. Isotachophoresis with carrier ampholyte mixtures (1) is agreed to be a quite different technique; as it is a pH gradient focusing method (1, 2), with different characteristics, it should not share the name transphoresis.

It is of interest that the use of the technique to separate isotopes, as mentioned above by Dr. Peel, was suggested by Lindemann (3) in 1921, independently of Kendall. The method was successfully used preparatively, with solvent counter-flow, to enrich isotopes of potassium, copper, and chloride (attempted by Kendall (4)) in the Manhattan Project (5); this is an interesting frontal variant of transphoresis in that the terminator ion is replaced by extension of the separand mixture to the terminator electrode. This work was extended in Germany (6) and elsewhere. More recently the field has been reviewed (7 - 10). One effect of the use of counter-flow is to increase the effective counter-ion transfer number at the expense of that of the separand ion mixture. Transphoresis in non-aqueous solvents has been achieved (14, 15).

The major breakthrough in the analytical use of transphoresis was made by Martin, who realised the significance of Kendall's work (11), and introduced thin-walled capillaries in 1944-1946 (12, 13), on which he separated small organic anions, using thermal detection. It was at Martin's suggestion that Longsworth (12) in 1953 used Kendall's method to separate various ions, including amino-acids, in free solution with counter-flow, in a Tiselius cell with Schlieren optics. The first protein separations by transphoresis in tubes were done in 1969 (16), including a serum transpherogram. We also observed at the time the sieving effect of various concentrations of non-gel polymer solutions, and found we could thereby raise the step height of serum albumen until it was overtaken by the glycinate terminator. Step order may be changed by use of sieving effects. Continued work to pioneer the commercial development of analytical transphoresis is now being done by my group, aimed at achieving fast automatic analysis with gradients in excess of 1 kV/cm. A recent analysis of 100 p-eq of calcium is by no means the limit of sensitivity.

Criticism has been made of external thermal detection, and deserves comment. The sluggish response with such detection is partly due to the thermal capacity of the tube wall and contained solvent, the latter being less easily reducible. Attainment of equilibrium step height resembles an inverted cooling curve whose period is reducible by improved cooling, which in turn reduces the already small signal, with increased extraneous signal and noise problems. The poor cooling required for manageable signals also results in longitudinal temperature gradients, with attendant chemical and pH gradients and other anti-separative influences. Moreover, with free convection Newton's law of cooling does not apply, so step height is not a linear measure of resistance. In 1970 I used switched current changes to co-record maximal thermal response speed and the faster response of an internal electrode pair system – one of several universal electrical detection methods I have used subsequently for fast analysis with direct cooling. This is important, as speed and resolution of detection are at present the main limitation of speed and precision of transphoretic analysis, fast detection opening the door to automation.

Refs. 1, 2, 13 & 16 *correspond respectively to* 1, 30, 9 & 13 *in the preceding contribution by Dr. Peel.*

3. Lindemann, F.A., *Proc. Roy. Soc. (Lond.) A 99* (1921) 102.
4. Kendall, J. & White, J.F., *Proc. Nat. Acad. Sci. (Wash.) 10* (1924) 458.
5. Brewer, A.K., Madorsky, S.L., *et al.*, *J. Res. Nat. Bur. Stds. U.S.A. 38* (1947) 137.
6. Martin, H., *Z. Naturforsch. 4a* (1949) 28.
7. Cole, H.C. & London, H. in *Separation of Isotopes* (H. London, ed.) G. Newnes, London (1961) p. 387.
8. Gazith, M. & Roy, A., *Electrochem. Technol. 2* (3 - 4) (1964) 85.
9. Freyer, H.D. & Wagener, K., *Agnew. Chem.* Internat. Edit. in English *6* (9) (1967) 757.
10. Preetz, W., *Fortschr. Chem. Forsch. 11* (3) (1969) 375.
11. Consden, R., Gordon, A.H. & Martin, A.J.P., *Biochem. J. 40* (1946) 33.
12. Longsworth, L.G., *Circ. U.S. Natl. Mus. 524* (1953) 59.
14. Preetz, W. & Pfeiffer, H.L., *J. Chromatog. 41* (1969) 500.
15. Blasius, E. & Wenzel, U., *J. Chromatog. 49* (1970) 527.

Views on nomenclature (see also the above remarks on the term *Transphoresis*)

A.J.P. Martin originally proposed the term *Displacement electrophoresis*. The term *Carrier-free electrophoresis* has also been proposed (J. St.L. Philpot). The more recent term *Isotachophoresis* does have some rationale (J.St.L. Philpot; D. Peel). However, terms such as the latter, virtually invented at a late stage for commercial reasons, are to be strongly deprecated, and the established term *Displacement electrophoresis* should remain the preferred term (C.J.O.R. Morris; S. Hjertén).

Fractionation of serum proteins by isotachophoresis

Terry Rosenbaum *(LKB Produkter AB, Sweden)*

With an LKB Uniphor electrophoresis apparatus, human serum proteins may be fractionated on a preparative scale. The elution of the protein zones from the bottom of the acrylamide gel column can be accomplished by the use of the same Tris-cacodylate buffer as that used for the leading buffer Ampholine carrier electrolytes pH 5-7 are used to space the protein zones. Nine well defined protein fractions, identified by immunoelectrophoresis, are obtainable.

Comments on these serum protein studies (J.O.N. Hinckley).- The need for spacers and subsequent zone elution procedures may reflect the inadequacy of UV as a detection system in transphoresis. I was able to record arguably nine components from serum by transphoresis in free solution with Tris as buffer counter-ion, using a glycine terminator and a 15 mM chloride leader. This was done in a stationary PTFE capillary cooled by freely convecting air, using as poor a detector as the external thermocouple. This serum separation (as cited above) is, I believe, the first to have been performed by transphoresis.

Other comments on displacement electrophoresis.- Selection of the proper mobility for the counter-ion is very important (T. Rosenbaum). The technique is especially applicable to small ions (D. Peel, *answering* J.O.N. Hinckley). It is doubtful whether moving boundary theory is strictly applicable to proteins (M.P. Tombs; C.J.O.R. Morris).

CHROMATOGRAPHIC METHODS

New chromatographic materials *(cf. Article 10)*

A.R. Thomson (*replying to* S. Hjertén, M.K. Joustra, C.J.O.R. Morris, A. Pusztai and T. Rosenbaum).- The bead materials are prepared by a sintering process; they are due to come on the market, and should not be unduly expensive, particularly since the starting materials (titania and calcium phosphate) cost only around 50p/lb in bulk. With hydroxyapatite, phosphate and pyrophosphate buffers are recommended, rather than citrate. Citrate elutes effectively from titania; NaCl is ineffective. For some enzymes subjected to bead chromatography, e.g. ribonuclease, there is almost complete recovery; with PGK the recovery of activity is 65-70% after chromatography on titania - which is regarded as satisfactory particularly since the enzyme is somewhat unstable. Concerning the presence in the illustrative chromatograms of about 10-30% protein which is eluted finally with 0.1 N NaOH, elution of tightly adsorbed protein can in fact be achieved with 2 M phosphate buffer; NaOH was used merely for convenience to make the column ready for use again.

The fractionation of living cells on DEAE-cellulose columns: the isolation of a parasitic protozoan genus *(Trypanosoma)* from infected host blood

Sheila M. Lanham *(Trypanosomiasis Res. Group, Lister Inst., London)*

African trypanosomiasis is a disease of man ('sleeping sickness') and domestic stock caused by a protozoan parasite, the trypanosome, which is transmitted by the tsetse fly. The organism is a highly motile flagellate about 1.0 µm in width and 10-30 µm in length. The bloodstream forms of the African trypanosome species cannot be cultured *in vitro*, and material for research must be harvested from the blood of the infected host.

Until recently, it was difficult and laborious to obtain trypanosome suspensions free of host cells from large volumes of infected blood. A method of separation, using short wide columns of DEAE-cellulose (Whatman Chromedia, DE 52), has now been developed (1, 2) which exploits the difference in negative surface charge, at pH 8.0, between the blood cells of the host and the infecting trypanosomes. The ionic strength of the equilibrating buffer was adjusted so that complete adsorption onto the DEAE-cellulose of all the negatively charged blood cells was just attained whilst the less negatively charged trypanosomes passed out with the eluate. The trypanosomes were recovered and washed free of plasma by centrifugation. High yields of viable, infective trypanosomes were obtained, uncontaminated with blood cells and platelets. Blood samples from 0.1 to 200 ml have been processed. The correct adsorption conditions were determined for a wide range of erythrocyte species, and with a few exceptions the adsorption behaviour corresponded to the degree of negative surface charge. Trypanosomes were recovered from blood containing as few as ten organisms per ml, and the diagnostic use of the method is under investigation. The method has also been successfully applied to the separation of spirochaetes from infected blood (3).

The relative surface charge of the various trypanosome species was established by adsorption-elution experiments on DEAE-cellulose. With one exception the series corresponds to taxonomic status.

1. Lanham, S.M., *Nat. (Lond.) 218* (1968) 1273.
2. Lanham, S.M. & Godfrey, D.G., *Exptl. Parasit. 28* (1970) 521.
3. Ginger, C.D., *Trans. Roy. Soc. Trop. Med. Hyg. 28* (1970) 357.

Note added by Editor.- For immunocompetent lymphocytes an affinity chromatography procedure may be employed, with bound antigen. Thus, use has been made of glass or plastic beads (1), Sepharose (2), polyacrylamide (3) and various plastics (4).

1. Wigzell, H. & Andersson, B., *Ann. Rev. Microbiol. 25* (1971) 291.
2. Davies, J.M. & Paul, W.E., *Cellular immunol. 1* (1970) 404.
3. Truffa-Bachi, P. & Wofsy, L., *Proc. Natl. Acad.Sci. (Wash.) 66* (1970) 685.
4. Edelman, G.M., Rutishauser, U. & Millette, C.F., *Proc. Natl. Acad. Sci. (Wash.) 68* (1971) 2153.

New aspects of gel filtration

L. Fischer (Pharmacia AB, Sweden)

The methodology of gel filtration in laboratory columns was, at an early stage, developed to a satisfactory degree for most purposes, and has since then remained largely unchanged. The recent development of high pressure chromatography has not influenced gel filtration because the gels are so soft that the beds are rapidly compressed under the pressures used. The main developments in recent years have therefore been in the extension of the method to new fields of application. Methodologically the most important developments have been the extension of the scale of the method downwards and upwards.

The extension downwards has been achieved by the development of thin layer gel filtration (TLG). Earlier attempts were made to use chromatographic gels for thin layer work (1). It was not, however, until Johansson and Rymo (2) developed the moist layer technique that thin layer gel filtration could be used practically. The reason is that when liquid penetrates into a dry layer of chromatographic gel the particles swell. As they do so they take up liquid which is taken from the layer behind the front. The flow rate in this layer is therefore higher than the rate at which the front advances, and most substances in a dry layer experiment would be carried to the front. If the layer is moist, liquid must be carried through the layer by gravity, and the experimental set-up therefore differs from that in TLC. In its simplest form TLG is carried out with the layer spread on the glass plate with a lid fastened a few mm above the plate. The gel layer is connected to eluant reservoirs by filter paper bridges at the ends. The flow is produced by a difference in level between the eluant reservoirs and can be controlled by regulating the levels.

This set-up is simple but difficult to handle reproducibly. A more suitable apparatus for routine TLG has been developed in the laboratories of Pharmacia, with a chamber which incorporates the eluant reservoirs.* With a TLC spreader or with a stainless steel rod terminated by distance pieces (Stahl), the layer (0.4-1 mm) is spread by moving an excess of thick gel suspension back and forth until an even layer had been formed. Excess gel is pushed off the plate. After a settling period, the sample (5-10 ul if a spot) is applied with the plate still horizontal. The chromatogram is developed by inclining the chamber. The best staining method is to make a replica by placing a dry filter paper on top of the gel layer for 1 min, the paper then being left to dry horizontally and stained as in paper chromatography.

TLG will probably be most interesting for clinical chemists in the diagnosis of paraproteinaemias and for molecular weight estimation. For the latter application the method has the great advantages that is requires very little substance, is insensitive to the composition of the eluant and does

* *The author's account of the apparatus has been shortened.- Editor*

not require the substance to be pure. From column chromatography experiments it is known that within the fractionation range of the gel type a linear relationship is obtained between log M and V_e. In TLG this corresponds to a linear relationship between log M and the inverse of the distances migrated.

The extension upwards of the scale of gel filtration is represented by the development of columns and equipment for industrial gel filtration, e.g. for desalting as practised in the pharmaceutical and food industries.- No regeneration is required as long as the gel bed is not contaminated, and it has very high capacity. The techniques used for desalting where large columns are used cannot be applied to gels that are intended for fractionation of protein mixtures etc., because the gels are not sufficiently rigid. With soft gel types we have found that the most practical way of making preparative columns is to use very short and wide columns. To obtain a sufficient bed length, several columns must be connected in series, each unit being typically of 37 cm diameter, 15 cm bed height, 16 l bed volume. By constructing the top and bottom pieces in a special way, zone broadening due to the connections is reduced to a level that is negligible.

Various protein mixtures run in 5-7 such sectional columns have shown resolution comparable to that obtained with analytical columns in the laboratory. Sectional columns will probably be most useful for the pharmaceutical industry. Processes have been developed for separation of insulin from pro-insulin (big insulin), for purification of albumin and transferrin from Cohn fractions, and for fractionation of pure serum (3). Sectional columns may also be used for chromatography on Sephadex ion-exchangers, one application of this being the preparation of IgG (4).

1. Determann, H., *Experientia 18* (1962) 430.
2. Johansson, G. & Rymo, L., *Acta Chem. Scand. 16* (1962) 2067.
3. Jansson, J.-C., *J. Agr. Food Chem. 19* (1971) 581.
4. Joustra, M. & Lundgren, H., *Protides of the Biological Fluids 17* (1970) 511.

L. Fischer (*answering queries by* H.K. Robinson, T. Rosenbaum and others).- With certain proteins including haemoglobin and possibly asparaginase, slight adsorption on to Sephadex commonly occurs, such that they are retarded more than would be expected from their known molecular weight and shape. In the TLG procedure there is little diffusion of separated proteins if the paper replica is prepared promptly. When Sephadex columns are subjected to very high pressures to restore the flow rate to a high level, reproducibility is said to be not very good. In the large-scale column work with Sephadex G-200 the flow rate is about 3-4 $ml/cm^2/h$.

Query (D.H. Leaback).- In view of the proven advantages of using beads of uniform size and shape in other branches of chromatography, what is known concerning the effect of such factors on the flow rates and resolving power for proteins in gel permeation chromatography, and what efforts do the manufacturers make to produce and control beads for uniform size and shape ? *Answer.*- There are publications indicating that there are advantages in uniformity of bead size and shape, but I doubt whether this is particularly critical in the gel permeation of proteins. The cost of further grading of Sephadex beads by the manufacturer could be prohibitive unless there were great practical advantages in doing so.

METHODS WITH AN ORGANIC PHASE

Nucleic acid preparations separated by use of phenol *(cf. Article 14)*

Concerning the properties of components thus separated (R. Williamson, *answering questions by* E.H. Hartley and R.H. Hinton), the capacity of poly-A to bind to membrane filters has been made use of in its assay, as has specific hybridisation to poly-dT bound to membrane filters (refs. 6-9 in the article); gel electrophoresis does help resolve the identity of functional RNP complexes, although RNP migrates in polyacrylamide gels to approximately the same position as the RNA species it contains.- The reasons have not been fully explored, and further research into this behaviour is needed.

Phenol methods for proteins *(cf. Article 16)*

Note by E. Reid.- Advantage has been taken (1) of the solubility of pituitary growth hormone in phenol and certain other solvents (containing up to 5% water) to attempt fractional extraction from a dry column containing the hormone preparation mixed with Supercel; the solvent admixed with a non-polar solvent such as ether served as eluant. Despite some denaturation, there was some recovery of active hormone, albeit not more active than the starting material.

1. Reid, E. & Wilhelmi, A.E., *Proc. Soc. Exp. Biol. Med., N.Y., 91* (1956) 267.

ISOPYCNIC AND SEDIMENTATION METHODS *(cf. Articles 18-20)*

G.B. Cline has formed a favourable opinion of the efficacy of 'reograd' (reorienting gradient) rotors as compared with ordinary zonal rotors that are loaded and unloaded whilst in motion.

Concerning the use of zonal rotors to separate cells, see Vols. 1 & 3 in this Series.- Editor.

On the question of the mechanism that determines at what interface cells settle when sedimenting through a discontinuous gradient (K.A. Dicke), the changes which occur in cells as they pass through such a gradient are not understood. It is obvious that cells behave differently when they are sedimenting through a discontinuous gradient than when they pass through a continuous gradient; they settle at different densities, yet the same general pattern is observed on the two gradients. Concerning the possible need to oxygenate the gradients in which lymphoid cells are separated (insofar as separation of liver cells requires that the sedimentation medium be adequately oxygenated), K.A. Dicke and the other contributors have not examined the need for oxygenation, but reckon that the low consumption of oxygen by lymphoid cells makes this unnecessary.

SYNCHRONIZATION OF DATA READ-OUT AND CUT NUMBERING IN SPECTROPHOTOMETRIC MEASUREMENTS ON EFFLUENTS

L. Funding, R. Larsen and J. Steensgaard *(Univ. of Aarhus, Denmark)*

Spectrophotometric measurements on effluents from chromatographic columns or zonal rotors (1, 2) for automatic computer processing have to be collected with precise connection between the data and the corresponding cut numbers. With most spectrophotometers and fraction collectors this presents a problem, as such items of equipment normally have no devices for synchronization.* To solve this problem we have constructed the following very simple device which lets the spectrophotometer control the tube changing. It is designed for use with a Gilford 2400 spectrophotometer and an LKB fraction collector. The principle is that the timing circuit which controls the cuvette changing is extended to control the tube changing too.

A relay is connected to the circuit of cuvette indicator lamp No. 1 (Fig. 1). The relay is mounted as a switch in the connections to the drop/siphon plug in the fraction collector. Hence every indicator lamp cycle is counted as one drop. Induction currents backwards from the relay must be counteracted by a diode.

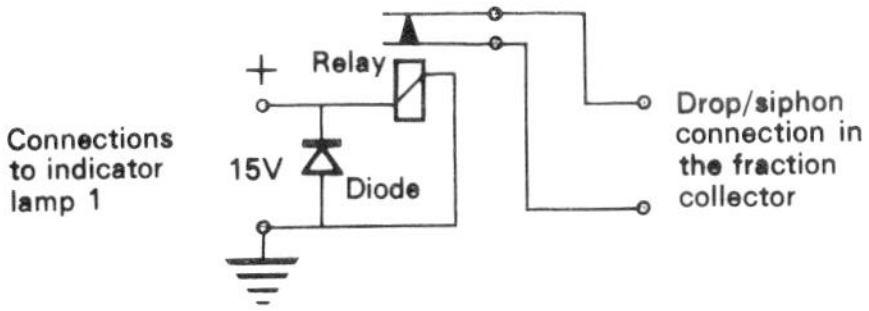

Fig. 1. Circuit diagram of the synchronizing device.

By use of this device the tubes will be changed after a pre-set number of measurements at one or two wavelengths. With a constant flow through the cuvette the measurements for each cut may be averaged. This way of correlating spectrophotometric measurements and cut numbering has proved convenient and satisfactory in several experiments in our laboratory.

1. Steensgaard, J., *Eur. J. Biochem.*, *16* (1970) 66.

2. Steensgaard, J. & Funding, L., *Acta Path. Microbiol. Scand. B 79* (1971) 19.

* *There are, of course, fraction collectors (e.g. the 'Central') with a plug-in for event marking on the spectrophotometer trace (cf. p.A-18, Vol.1)-Ed.*

EXPERIENCES WITH THE GeMSAEC FAST ANALYZER

R.S. Atherton *(A.E.R.E., Harwell, U.K.)*

The GeMSAEC fast analyzer was developed by Anderson (1). Samples and reagents are centrifugally transported into a multiple-cuvette rotor (15 to 42 cuvettes) spinning between a stationary light-beam and a photomultiplier. All cuvettes fill simultaneously on acceleration of the rotor. Wavelength selection can be by monochromator or interference filter.

We were much attracted to this concept since it seemed the only method which could give sufficiently rapid assays on the many fractions resulting from our zonal centrifugation, preparative electrophoresis (1) and chromatographic (3) projects. We therefore obtained a prototype instrument (M.S.E. Ltd.). This has a 16-place rotor operating at 500 rev/min, hence each cuvette may be read eight times per second if desired. The transmission through every cuvette is continuously displayed on an oscilloscope screen. The photomultiplier impulses are also transmitted via a digital voltmeter to a fast tape punch. Observation of the oscilloscope screen enables fractions containing a given enzyme to be identified semi-quantitatively in a matter of seconds, or in about 2 min including the sample loading. This is of great importance when continuous-flow preparative techniques are under study. We can easily assay 100 fractions per hour.

The enzymes which we assay routinely mostly result in the oxidation of NADH to NAD^+ and include lactate dehydrogenase, phosphoglycerate kinase and myokinase. We also assay protein by both the Biuret and Folin procedures. In principle any assay that can be carried out in a spectrophotometer can also be carried out in GeMSAEC. Since the cuvette volume is only 0.6 ml this gives a great saving in reagents. Since readings are taken at such frequent intervals and blanks and standards can be analysed simultaneously, end point reactions need not be carried to completion, and accurate data for the determination of initial reaction rates can be obtained in only one minute, or even less in many cases.

Acknowledgement

This note appears with the sanction of the Atomic Energy Research Establishment.

1. Anderson, N.G., *Anal. Biochem. 28* (1969) 545.
2. Unpublished work at A.E.R.E. Harwell.
3. Thomson, A.R., Miles, B.J., *this volume, Article 10.*

INDEX of Subjects

Since the book is concerned with methods, particular examples of materials separated (and their sources) are usually NOT indexed; assay procedures for examining products are likewise excluded. Authors of the main contributions are listed on p. 8.

Page entries such as 175- signify that ensuing pages are also relevant, *i.e. the - connotes a major entry.*